Programming and Algorithms:

An Introduction

Programming and Algorithms:

An Introduction

Anthony J. Guttmann
University of Newcastle, New South Wales, Australia

Heinemann Educational Books
London

Heinemann Educational Books Ltd

LONDON   EDINBURGH   MELBOURNE   AUCKLAND   TORONTO
HONG KONG   SINGAPORE   KUALA LUMPUR   NEW DELHI
NAIROBI   JOHANNESBURG   LUSAKA   IBADAN
KINGSTON

ISBN 0 435 77541 3

Published by Heinemann Educational Books Limited
48 Charles Street, London W1X 8AH

Text set in 10/12 pt IBM Press Roman, printed by photolithography,
and bound in Great Britain at The Pitman Press, Bath

To the memory of my parents.

Preface

This book is based on an introductory programming course of thirty lectures that has been given at the University of Newcastle, New South Wales, Australia for the past five years. It discusses and develops the contemporary view of computer programming and algorithm design. Unlike most other introductory programming texts, some prior knowledge and experience with a programming language is assumed. This approach recognizes a growing educational trend in which students receive some elementary programming instruction (usually in BASIC, FORTRAN or ALGOL) at a very early stage — often as part of a first year university or similar mathematics course, or even at school.

The first chapter develops the ideas of top-down program design, structured programming, and program style. Six basic control structures are introduced and their role in writing well-styled, well-structured programs is fully discussed. The three languages used to represent a program in the planning stage: natural language, decision tables, and flowcharts, are also introduced at this stage.

The next three chapters are devoted to a discussion of the three most commonly used scientific programming languages: BASIC, FORTRAN, and ALGOL. I do not share the view that since BASIC and FORTRAN are poor languages in which to write well-structured programs, they should be avoided. Rather, I take the view that since these languages are widely used, and are likely to remain so for some time, consideration must also be given to writing well-styled, modular programs in these languages. The language features of ALGOL on the other hand allow the ideas of structured programming to be developed much more fully.

In Chapter 5 the two languages FORTRAN and ALGOL are discussed and compared. This comparison leads to a discussion of recursive procedures and a way of simulating these in FORTRAN is presented. The question of program planning and design forms the subject of the next chapter, with special emphasis being placed on the most efficient methods of coding various common problems. In Chapter 7 the subject of algorithm design and analysis is considered. It is this aspect that is probably the most creative part of program design, and some guidelines for the design and choice of suitable algorithm. are given.

The remainder of the book continues the study of algorithms by considering a number of numerical and non-numerical problems. Chapter 8 is concerned with sorting and searching algorithms, while Chapter 9 is devoted to simulation problems, and includes a review of elementary statistics and the theory of random number generators.

At the end of each chapter a number of problems are given, most of which are designed to illustrate the material in that chapter while a few are chosen to extend it. Solutions to most of the problems are given at the end of the book. A bibliography is also given at the end of the book, in which pertinent references for each chapter can be readily found.

Where possible, the examples presented in the text are machine independent, but where it is necessary to refer to a specific machine, I have tried to be quite general by referring to several, such as the IBM 360, 370, and 7000 series, the ICL 1900 series, the CDC 6000 series, and the PDP 11 series.

Students with only one year of university level mathematics should be able to cope with most of the material, but may find parts heavy going, notably parts of Chapters 5, 8, and 9. These sections may be skimmed on first reading. Students who have taken, or are currently taking, a second year mathematics course should have no difficulty with the level of the text.

Readers familiar with D. Knuth's *magnum opus* will recognize my indebtedness to him. It is a pleasure to thank E. A. Ashcroft, J. M. Blatt, D. C. Cooper, R. J. Dear, R. Freak, S. V. Guttmann, J. A. Lambert, D. L. S. McElwain, J. I. Munro, and C. J. Thompson for their comments on various sections of the manuscript, and R. J. Vaugha for the railway crossing example of Chapter 9.

Finally, I would like to thank Mrs K. Abraham and Mrs V. Herridge for their expert typing of the manuscript.

1977   A.J.G.

Contents

1. Top-down Programming

1.1 Introduction

In the early days of computer science storage was of paramount importance and programs were largely judged on the basis of their storage requirements and on whether they ran successfully or not. As computers became larger and more expensive, it became possible to write much longer and more complex programs than were written for the earlier, comparatively primitive, machines. As a result, the simple test of whether the program ran or not became more difficult to apply. Firstly, because the program could be so much longer, it was far more difficult to test all possible situations. Thus a program which appeared to be satisfactory could fail on a particular combination of data for some obscure reason. Secondly, since programs could be so much longer, it became increasingly necessary to consider questions of efficiency.

The concept of efficiency itself has changed considerably over the last decade. Ten years ago, any discussion of efficiency was primarily concerned with questions of processor time and storage requirements. While these are certainly not negligible considerations even today, the cost of computer hardware has rapidly decreased, while manpower costs have increased. Thus questions of efficiency now tend to focus on the minimization of manpower costs. As a result, efficiency now tends to encompass such concepts as program readability and maintainability. Programming courses should properly concentrate on stressing these aspects, as well as the methodology of writing programs in a manner that efficiently utilizes the programmer's time. Such a methodology should result not only in readable, maintainable programs, but programs that require a minimum of debugging. It is difficult to overstress this last point, since it has been estimated that as much as 50 per cent of the programmer's time has been spent on debugging programs.

In response to these changing requirements of program writing, a more systematic view of the process of computer programming has begun to emerge. The stage has not yet been reached — or even forseen — where a set of rigid rules can be mechanically

1

applied to any programming problems, resulting in a first-class program. Rather, certain guidelines can be given, which, if intelligently followed, should result in better programs being produced. The process is not unlike that of writing a piece of prose. There exist the rigid rules of the written language – the syntax – that must be obeyed, but in addition there are generally accepted notions of style which can greatly facilitate the production of a piece of prose. Nevertheless, following these stylistic guidelines is not sufficient to turn the budding author into a Shakespeare.

In this book the emphasis is on those aspects of program planning, style, and construction which should lead to the writing of better programs. The basic approach is called *top-down programming,* variously known as programming by successive refinement or hierarchical modularity. What is meant is that the program is first planned 'in-the-round', with a small number of distinct, highly-abstracted processes to be performed successively. Next, these processes are considered individually and each one is broken into a small number of distinct, less-abstracted processes. The process of working out successively finer detail – corresponding to reducing the level of abstraction – is continued until the solution to the problem is at a stage where it can be directly translated into the required programming language.

The antithesis to top-down programming is *bottom-up programming.* This was the style the professional programmer used without a second thought only a few years ago and it is a style that should be avoided today. To illustrate these two approaches, consider the problem of obtaining the roots of the quadratic equation

$$ax^2 + bx + c = 0$$

In the bottom-up approach, the programmer immediately realizes that he will require a statement to evaluate

$$(b^2 - 4.a.c)^{\frac{1}{2}}$$

when solving the equation. Assuming he is writing in FORTRAN, the first thing he writes down could be

$$DISC = SQRT(B*B - 4.0*A*C).$$

This statement is highly specific, and completely non-abstracted. It corresponds to the lowest level or 'bottom' of the program. From this original statement, bits and pieces are added at random as they occur to the programmer, and the innumerable errors that inevitably occur are 'patched up' as they are found. The result is a program that has taken a considerable time before running successfully, and yet is still of rather doubtful reliability because of the haphazard nature of its construction.

In the top-down approach, the programmer stops and thinks about the problem before writing anything down. He will consider such mundane questions as – 'Which programming language is appropriate to the problem? In what form will the data be presented? What special cases can occur?' – and will jot down notes to himself. Trivially, the problem breaks down into three phases, that of input, followed by the calculation phase, followed by the output phase. Each stage is considered in turn, and successively refined. The strategy can be illustrated by a 'tree', as shown in Figure 1.1.

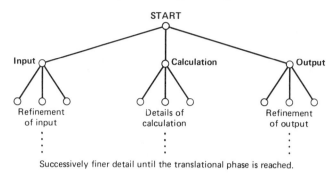

Successively finer detail until the translational phase is reached.

Figure 1.1

At each level the problem is less abstracted than at the previous level. Also, since there is no connection between various stages at the same level, if a change is needed at any point due to an oversight, or just because a better way of doing something has been found, one has only to work back up the tree to the level at which the change is needed. All higher levels remain unchanged. Further, a change in the calculation phase requires no change in the input or output phases due to the modular construction of the program, even at this early stage of the program's construction. In the bottom-up approach, *any* change requires careful, yet highly error-prone analysis of the entire program before the necessary changes can be made. This is because all parts of a program developed in a bottom-up manner can — and usually do — depend on all other parts.

Together with the top-down approach to program design, a great deal of attention has been given to program construction in recent years. As a result there now exists a methodology of program construction called *structured programming*. This can be thought of as an engineering approach to program construction, in the sense that it is planned and systematic in contrast to the *ad hoc* methods characteristic of the older, more intuitive method of programming. The term 'structured programming' was first introduced by Dijkstra, Dahl, Hoare, Wirth, and others. It originally meant 'structured' in the widest sense of the word, encompassing the form of program statements, data types, and program descriptions. Nowadays there is a tendency to restrict the term to the proper structuring of program statements, while an even narrower view essentially identifies structured programming with the elimination of unconditional transfers, that is, GO TO statements.

N. Wirth has recently given a definition of structured programming which reads in part: 'It is the expression of a conviction that the programmers' knowledge must not consist of a bag of tricks and trade secrets, but of a general intellectual ability to tackle programs systematically, and that particular techniques should be replaced (or augmented) by a method'.

I have attempted to keep to this viewpoint throughout the book but have laid the heaviest emphasis on one aspect, the use of a limited set of statement types that effect the communication between different parts of the program. Such statement types are called **control structures**. The use of these control structures, when augmented by the top-down approach to program design, should lead to well-structured programs.

Before considering any specific programming language, or even the set of control structures we will use, we must consider how to represent the solution of whatever programming problem is at hand in such a way that the top-down approach can be readily applied, and the translation to the desired programming language easily performed. This formulation of the program is usually carried out using either flowcharts, decision tables or natural language or a combination of these. We will study each of these in turn.

Flowcharting is a fairly well established practice and indeed many introductory programming courses commence with a study of flowcharting. Once a flowchart is written for a simple problem, it is usually fairly straightforward to translate the flowchart into a language such as FORTRAN or ALGOL.

Decision tables have been much vaunted in recent years, but still do not appear to have received the more wide-spread use they undoubtedly deserve. One possible reason for this is that the process of translating a decision table into say, a FORTRAN program is not as straightforward as is the translation from a flowchart. On the other hand, there do exist programming languages and techniques to handle problems formulated by decision tables, but these are not very widely known.

A *natural language* is simply a precise language using a subset of English for its expression in which a sequence of operations can be conveniently formulated. The statements used must be clear and unambiguous and are usually based upon a written form of the basic control structures referred to above.

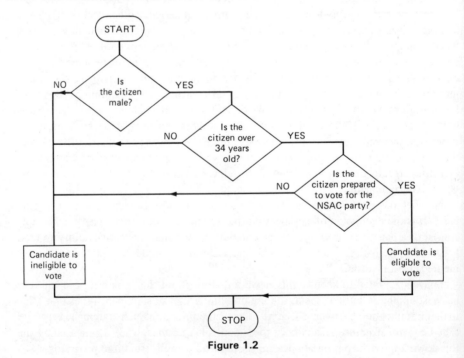

Figure 1.2

We will discuss a recommended set of control structures at the end of this chapter, but for the moment let us consider a simple programming problem using each of these methods.

In the country of Lower Slobovia, the ruling military junta, fearful of their position, are about to hold the first general election in 20 years as part of their democratization campaign.

Prior to the elections, a list of eligible voters is prepared. The following restrictions apply:

(1)   women may not vote;
(2)   only males over the age of 34 may vote;
(3)   only people who are prepared to vote for the ruling NSAC (National Socialist Anarchic Communist) party are eligible.

A flowchart to test the eligibility of a citizen could be as shown in Figure 1.2.
A decision table for the same problem might be the following:

Is the citizen male?	N	Y	Y	Y
Is the citizen over 34?	–	N	Y	Y
Will the citizen vote NSAC?	–	–	N	Y
The citizen is eligible to vote				X
The citizen is ineligible to vote	X	X	X	

Note that an X represents the action to be taken, given the conditions above.
The natural language approach would be to write the solution as follows:

> IF the citizen is male and is over 34 years old and is prepared to vote for the NSAC party THEN he should be entered on the eligible voters list ELSE not.

The representation of the solution is straightforward in all cases and would be readily understood by someone with no knowledge of these methods. Note that in all cases we have discarded irrelevant choices; for example, having established that a citizen is female, we don't test her age or voting behaviour, since these do not affect our decision. In the natural language approach however, once a condition following an IF is not satisfied, we skip straight to the statement following the ELSE. This is an eminently sensible convention, but it does need stating, as otherwise ambiguities in interpretation could arise.

Let us now look at each method in more detail.

1.2 Flowcharts

Since a major purpose of this stage of program preparation is that it should help a person understand a program written by another person, it is important that standard symbols be used to represent different operations. There is an internationally agreed standard for the flowchart symbols and the thirty standard symbols are shown in Figure 1.3. Many manufacturers make a template with all standard flowchart symbols on it. These are inexpensive and convenient, and the purchase of one is recommended.

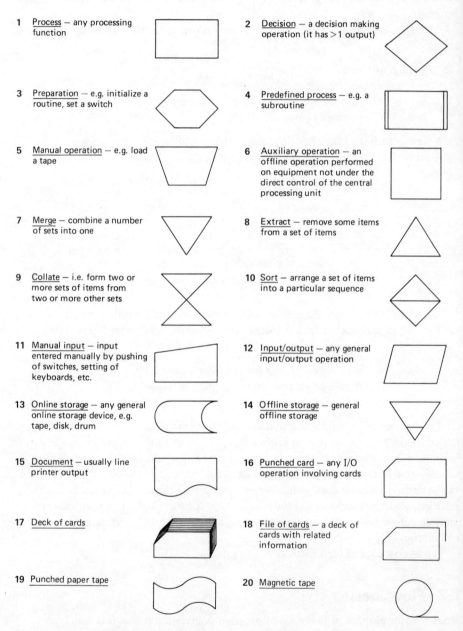

1 Process — any processing function

2 Decision — a decision making operation (it has >1 output)

3 Preparation — e.g. initialize a routine, set a switch

4 Predefined process — e.g. a subroutine

5 Manual operation — e.g. load a tape

6 Auxiliary operation — an offline operation performed on equipment not under the direct control of the central processing unit

7 Merge — combine a number of sets into one

8 Extract — remove some items from a set of items

9 Collate — i.e. form two or more sets of items from two or more other sets

10 Sort — arrange a set of items into a particular sequence

11 Manual input — input entered manually by pushing of switches, setting of keyboards, etc.

12 Input/output — any general input/output operation

13 Online storage — any general online storage device, e.g. tape, disk, drum

14 Offline storage — general offline storage

15 Document — usually line printer output

16 Punched card — any I/O operation involving cards

17 Deck of cards

18 File of cards — a deck of cards with related information

19 Punched paper tape

20 Magnetic tape

Figure 1.3 The thirty standard flowchart symbols

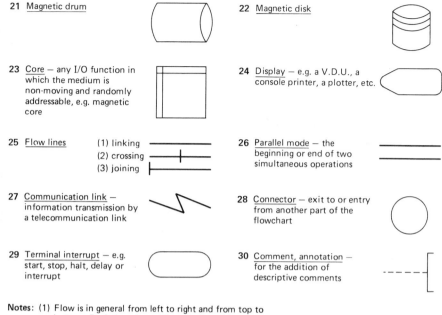

21 Magnetic drum

22 Magnetic disk

23 Core – any I/O function in which the medium is non-moving and randomly addressable, e.g. magnetic core

24 Display – e.g. a V.D.U., a console printer, a plotter, etc.

25 Flow lines   (1) linking   (2) crossing   (3) joining

26 Parallel mode – the beginning or end of two simultaneous operations

27 Communication link – information transmission by a telecommunication link

28 Connector – exit to or entry from another part of the flowchart

29 Terminal interrupt – e.g. start, stop, halt, delay or interrupt

30 Comment, annotation – for the addition of descriptive comments

Notes: (1) Flow is in general from left to right and from top to bottom. Deviations from this are indicated by open arrows. Crossed flow lines have no logical interrelation.

(2) Two or more incoming flow lines may join with one outgoing flow line.

Figure 1.3   (continued)

Let us look at some examples of flowcharting the solutions to a few numerical and non-numerical problems.

Example 1

The word *algorithm* has been traditionally associated with Euclid's algorithm for finding the greatest common divisor (g.c.d.) of two positive integers. (Note that the g.c.d. of *any* two integers m and n is equal to the g.c.d. of $|m|$ and $|n|$, so that Euclid's algorithm may be used to find the g.c.d. of any two integers, irrespective of sign.) The algorithm may be stated as follows:

(1)   Divide m by n and let r be the remainder, with $0 \leqslant r < n$.
(2)   If $r = 0$, the algorithm terminates and n is the g.c.d.
(3)   If $r \neq 0$, set $m \leftarrow n$ and $n \leftarrow r$ and return to step 1.

The flowchart for a program to find the g.c.d. of *any* two integers m and n using Euclid's algorithm is shown in Figure 1.4.

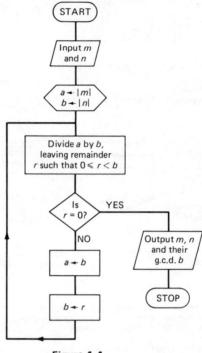

Figure 1.4

With a little experience, this pictorial representation is much clearer than the verbal description above. Indeed, one can follow the nature of the solution just from the shape of the boxes, without reading the information printed within them. In this case we see that after starting, an input operation takes place, followed by a preparation operation. A process then takes place, followed by a decision. If the decision goes one way, output is produced and the flowchart terminates. If the decision goes the other way, two further processes take place and control is transferred back to the initial process, and so on. Thus, from our knowledge of the standard flowchart symbols, we can follow the logic of the method of solution *without even knowing the problem*.

Example 2

Construct a flowchart for a program to calculate e^x using its power series expansion.

A value of x should be read in, and the series summed until the result is accurate to six significant figures. Output x and e^x, then stop.

$$e^X = 1 + X + \frac{X^2}{2!} + \frac{X^3}{3!} + \frac{X^4}{4!} + \ldots = \sum_{n=0}^{\infty} X^n/n!$$

We calculate successive terms by multiplying the previous term by X/n, and denote the estimate of the sum EX. Thus six figure accuracy will be attained when $|\text{TERM/EX}| \leqslant 10^{-6}$. The flowchart to implement this calculation is shown in Figure 1.5.

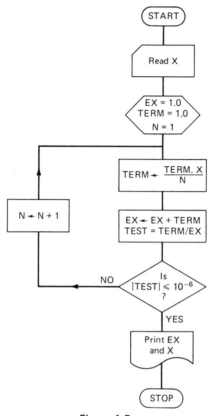

Figure 1.5

This example demonstrates another point that should be borne in mind. Though the method of solution and the flowchart are logically correct, translation of this flowchart to, say, ALGOL would produce a very bad program. Can you see why? Well if $X = -100$, then e^X is of course very small. However, the absolute value of the coefficients in the power series expansion will be very large for very many terms and rounding errors will swamp the solution. Overcoming this problem is set as an exercise at the end of the chapter. The point however must be considered that though a flowchart may be logically correct, it can still represent a poor solution to the problem at hand.

Example 3

The Crash-U-Quick Airline Corporation has announced a new pricing schedule for return fares from Sydney to Lower Slobovia. This fare structure is as follows:

The regular adult return flight, with no time constraint and unlimited stop-overs, is $A1000. With only one stopover, and a stay of between 30 and 100 days, the special excursion fare is $A750. A group of more than ten people travelling together can take the regular flight for $A850 each, while there is no group concession for the special excursion fare. However, adult passengers travelling for less than 30 or more than 100 days who require only one stopover may travel for $A850, or for $A750 in groups of more than ten. Finally, children under 14 years may travel for 50 per cent of the corresponding adult fare, while children under two years may travel for 10 per cent of the corresponding adult fare.

The flowchart for the implementation of this fare structure is shown in Figure 1.6. You should check it carefully to see that it corresponds to the description given above.

Note that the flowchart description of this non-numerical problem is by no means easier to follow than the verbal description. Compare this with the case for the two numerical examples discussed earlier, for which the flowchart did specify the nature of the solution much more clearly.

As a loose generalization it appears that flowcharts are better at representing solutions to numerical problems than non-numerical problems. Decision tables are usually more suited to the solution of non-numerical problems, as we shall now see.

1.3 Decision Tables

The basic structure of a decision table is a rectangle divided into four as shown.

Basic Layout of a Decision Table

Condition stub	*Condition body*
List of N relevant conditions	At most 2^N columns
Action stub	*Action body*
List of all possible actions	Particular action marked with an X

At top left is the condition stub, where the N relevant conditions are listed, one to each row. At top right is the condition body, with columns for each possible set of conditions, and if each has two possible states, then there are at most 2^N columns. At bottom left is the action stub, where the full range of possible actions are listed, one to each row, and at bottom right is the action body, with a cross in each column representing the appropriate action. Many authors distinguish between *limited entry* and *extended entry* decision tables — the distinction is not of great significance, however, though we will give an example of each.

Consider the airline fare structure problem we have just looked at as our last example. There are three possible age groups for a passenger, plus two possibilities for each of

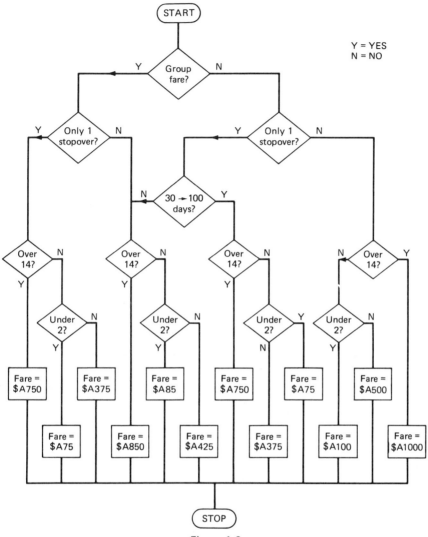

Figure 1.6

three further conditions; that is whether or not the passenger belongs to a charter group, whether or not there is more than one stopover, and the question of trip duration. Thus there are $3 \times 2^3 = 24$ possible sets of conditions; hence there will be 24 columns in the condition body. There are nine possible fares; hence there will be nine rows in the action stub.

The complete decision table is shown on page 12 and is an example of a *limited entry* decision table.

Note that certain columns in the condition body are left blank, since if a person is over 14 years of age he is certainly over two years of age. Also note that there are only

	1	2	3	4	5	6	7	8	9	10	11	12	13	14	15	16	17	18	19	20	21	22	23	24
Age $\geqslant$ 14	Y	Y	Y	Y	Y	Y	Y	Y	N	N	N	N	N	N	N	N	N	N	N	N	N	N	N	N
Age $\geqslant$ 2									Y	Y	Y	Y	Y	Y	Y	Y	N	N	N	N	N	N	N	N
Group > 10	Y	Y	Y	Y	N	N	N	N	Y	Y	Y	Y	N	N	N	N	Y	Y	Y	Y	N	N	N	N
1 stopover	Y	Y	N	N	Y	Y	N	N	Y	Y	N	N	Y	Y	N	N	Y	Y	N	N	Y	Y	N	N
Between 30 and 100 days	Y	N	Y	N	Y	N	Y	N	Y	N	Y	N	Y	N	Y	N	Y	N	Y	N	Y	N	Y	N
FARE = $A1000								X																
FARE = $A850				X																				
FARE = $A750		X				X																		
FARE = $A500	X		X		X		X																	
FARE = $A425										X		X		X		X								
FARE = $A375											X		X		X									
FARE = $A100									X															
FARE = $A85																		X		X		X		X
FARE = $A75																	X		X		X		X	

24 possibilities since we naturally do not allow the impossible case of someone being over 14 and under 2.

The use of this decision table is quite straightforward; for example, an adult passenger travelling alone for a trip of 70 days' duration requiring several stopovers corresponds to column 7 in the condition body, and thus to a fare of $A1000 as shown on the action stub.

We can simplify this decision table, however, by noting that columns 1 and 2, 3 and 4, 7 and 8, etc. correspond to the same fare, and hence to the same entry in the action stub. A look at the decision table shows that for these columns the trip duration is of no significance. A condition of no significance is represented by a dash in the appropriate column, and a simplified decision table is shown below. This decision table is an example of an *extended entry* decision table, with the three possible age groups shown as headings in the condition body rather than as rows in the condition stub.

Age is	Over 14					2 and over, yet under 14					Under 2				
1 stopover	N	N	Y	Y	Y	N	N	Y	Y	Y	N	N	Y	Y	Y
Group fare	N	Y	Y	N	N	N	Y	Y	N	N	N	Y	Y	N	N
Between 30 and 100 days	–	–	–	Y	N	–	–	–	Y	N	–	–	–	Y	N
FARE = $A1000	X														
FARE = $A850		X			X										
FARE = $A750			X	X											
FARE = $A500						X									
FARE = $A425							X			X					
FARE = $A375								X	X						
FARE = $A100											X				
FARE = $A85												X			X
FARE = $A75													X	X	

The passenger alluded to above — adult, travelling alone, several stopovers, and for 70 days — now corresponds to the first column, so that the fare is given correctly as $A1000.

Note that in this simplified decision table there are now only 15 columns in three groups of five rather than 24 columns as before. How do we know that we have allowed for all possibilities, or in other words, is the decision table still complete? This can be tested as follows: Each dash signifies that a particular two-fold choice is irrelevant; hence the two-fold choice is represented by a single column. Thus each dash represents one deleted column. There are nine such dashes, which, added to the 15 columns, gives us 24. This, of course, is the total number of possibilities, so the decision table is complete.

Sometimes we can have two or more crosses in the action body in the same column. This corresponds to two or more actions taking place in sequence, the sequence being read from top to bottom on the action stub. Note also that steps in the condition stub can be numbered, so that transfers can take place to that step. Both these points are illustrated in the following example.

Example

Draw a decision table to implement Euclid's algorithm, which was flowcharted earlier.

	N	Y	Y
1. HAVE N AND M BEEN READ?	N	Y	Y
2. IS $R = 0$?		Y	N
READ N AND M	X		
SET $A = \|N\|$	X		
SET $B = \|M\|$	X		
REPLACE A BY B			X
REPLACE B BY R			X
DIVIDE A BY B, CALL REMAINDER R, $0 \leqslant R < B$	X		X
GO TO 2	X		X
PRINT M, N, AND THEIR G.C.D. B.		X	
STOP		X	

The best way to see how this table works is to try it out with a specific pair of numbers, say 63 and 15, which have a g.c.d. of 3.

Finally, a delightful example of a decision table given by E. Humby. An ancient Persian poem goes:

> *He who knows not and knows not that he knows not*
> *is a fool – shun him.*
> *He who knows not and knows that he knows not*
> *can be taught – teach him.*
> *He who knows and knows not that he knows is*
> *asleep – wake him.*
> *He who knows and knows that he knows is*
> *a prophet – follow him.*

This can be written as a decision table as follows:

1. He knows.	N	N	Y	Y
2. He knows 1.	N	Y	N	Y
Shun him	X			
Teach him		X		
Wake him			X	
Follow him				X

One important topic we have not discussed is the translation of a decision table to a program. This is not as straightforward as the translation from a flowchart to a computer program and will not be discussed here. It is the subject of Humby's book, and is also discussed by Pollack et al.

1.4 Natural Language Formulation

One method of developing and expressing solutions to programming problems which is rapidly growing in popularity is the natural language method. In this method, the program is developed in stages, with the various operations and transfers of control written in an English-like language. There is not the same degree of standardization in this approach as there is in the flowchart and decision table methods. The programmer is free to write what he likes provided that it is clear and unambiguous. It need not be too specific, but if it bears some relation to an existing programming language, then translation into that language is straight-forward. For this reason many people tend to use an ALGOL-like language (frequently called 'pidgin ALGOL') to formulate their solutions. This ALGOL-like dialect is particularly appropriate for writing control structures, as can be seen by expressing our earlier examples in this language.

Example 1 was Euclid's algorithm, which could be written in natural language as follows:

> Input m and n; $a \leftarrow |m|$; $b \leftarrow |n|$;
> REPEAT BEGIN divide a by b leaving remainder r such that $0 \leqslant r < b$
> $\qquad$ IF $r \neq 0$ THEN
> $\qquad\qquad\qquad$ BEGIN $a \leftarrow b$;
> $\qquad\qquad\qquad\qquad$ $b \leftarrow r$
> $\qquad\qquad$ END
> $\quad$ END
> UNTIL $r = 0$;
> Print m,n and b as the greatest common divisor

Notice that this is very close to the verbal statement of the algorithm in the section on flowcharts. Clearly, one of the advantages of this method is that it is a more natural language in which to discuss problems than either flowcharts or decision tables. As the above example shows, it has the additional advantage of conciseness of expression.

The calculation of e^x (Example 2) in a natural language formulation follows the above lines and is left as an exercise for the reader. The airline fare-scheduling problem (Example 3) can clearly be expressed by a family of IF-THEN-ELSE statements of somewhat tedious length, as shown (in part) below:

> IF the customer is in a group of more than 10, requires only
> one stopover and is over 14 years old THEN the fare is \$A750
> ELSE IF the customer is in a group of more than 10 and only
> requires one stopover and is under 2 years old THEN the fare
> is \$A75 ELSE IF . . . etc.

Finally, the Persian poem can be considered to be already written in a natural algorithmic language — though unfortunately not in a style favoured by many programmers today!

We shall now study the control structures used in this method in more detail.
We shall standardize on a small set of six such structures. Such a set is small enough to be easily remembered, yet wide enough to express concisely most of the transfers of control we shall require.

1.5 Control Structures

Concatenation

The most obvious requirement in any programming language is the ability to string statements together sequentially. This is called *concatenation*, and is achieved in natural language by simply writing the statements one after another. So to increment x by 3, then to form $y = x + z$, we would write

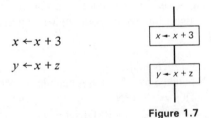

$$x \leftarrow x + 3$$
$$y \leftarrow x + z$$

Figure 1.7

where the left pointing arrow is to be read as 'is replaced by', the flowchart representation of this structure is shown in Figure 1.7.

Conditional Clauses

Often we wish to perform some operation if some condition holds, and to skip that operation otherwise. We write

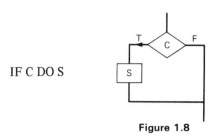

IF C DO S

Figure 1.8

where statement(s) S are to be carried out if condition C is true, but not otherwise. In flowchart form we have the situation shown in Figure 1.8.

Alternative Clauses

This is an obvious extension of the conditional clause to the case where we execute statement(s) S_1 if some condition holds and statement(s) S_2 otherwise. This control structure is written

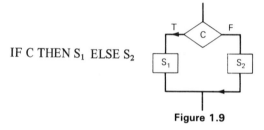

IF C THEN S_1 ELSE S_2

Figure 1.9

and is shown in flowchart form in Figure 1.9.

Choice Clauses

The situation frequently arises that we wish to execute one of a number of statement(s) depending on some condition. Thus we write CASE (i) OF $S_1, S_2, \ldots, S_N$ to mean that statement(s) S_1 is to be executed if i = 1, statement(s) S_2 is to be executed in case(2), and so on. The flowchart (Figure 1.10) is in this case a direct generalization of the two flowcharts above.

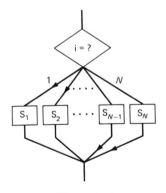

Figure 1.10

Repetition and Iteration

The distinction between these two cases is relatively minor. In *repetition* we repeat some operation *until* a condition is satisfied (testing after the operation), while in *iteration* we repeat some operation *while* a condition is true, testing before the operation. Thus we write

WHILE C DO S for iteration, and
REPEAT S UNTIL C for repetition.

The flowcharts for these two operations are shown in Figure 1.11.

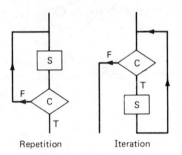

Repetition   Iteration

Figure 1.11

Frequently, iteration is achieved by simply incrementing a counter. This situation is so common that most languages provide a special control structure for that case, called a *loop*. This is written in natural language as shown below, and in flowchart form as shown in Figure 1.12.

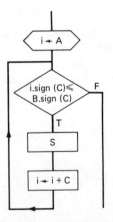

Figure 1.12

This means that i is set equal to A, and if i $\leq$ B then S is executed. i is then increased by an amount C, and if i.sign(C) $\leq$ B.sign(C), S is again executed. This process continues until i.sign(C) $>$ B.sign(C), whereupon S is skipped, and the next instruction is executed.

Unconditional Transfer

This is written GO TO L, where L is a statement label, and is shown in flowchart form by a line going to the appropriate statement. The role of this statement in the list of control structures has been the subject of much heated debate. This is because a notable characteristic of unstructured programs built from the bottom up is a bewildering array of GO TO statements, making the flow of control terribly difficult to follow. A corollary is that a well-structured program is likely to have few if any GO TO statements, though this feature alone is not sufficient to guarantee that a program is well structured. For this reason, the GO TO statement should be used infrequently and with care. Its proper application is to escape from a loop if some error condition is found, or to transfer to some 'end of program' or 'end of section' statement. In an unstructured language like BASIC or FORTRAN it must be used more often than this, due to the limited control structures directly available in these languages. Even then it should be used to transfer over as few statements as possible, and preferably only in a downward direction in the instruction sequence. In this way the program remains relatively readable, and can still be composed of discrete sections or **modules**.

Throughout this section, in referring to statement(s) S, we mean that S can be a simple statement, or several simple statements concatenated together, or even a group of statements containing control structures. Whenever S is more than a simple statement, it is a useful convention to indicate the beginning and end of the statement by statement parentheses BEGIN and END. For maximum clarity when writing in this natural language, both upper and lower case alphabets should be freely used. Both these points are illustrated in the solution to problem 1.7.

This completes our list of control structures. The use of this set should enable most programs to be written compactly, and will enable them to be designed in a top-down manner and translated into well-structured programs.

In the next few chapters we shall study the use of these control structures when writing programs in the three most commonly used scientific programming languages, which are FORTRAN, ALGOL, and BASIC.

Problems

1.1 Modify the flowchart for the calculation of e^X so that large positive and negative values of X may be accommodated, by removing the integral part of the exponent M and calculating e^M by multiplication separately.

1.2 Write a flowchart for the solution to the following problem. A card is to be read containing a number N. Following this card are N cards, numbered $1-N$, each of which contains a salary in whole dollars within the range \$A0–\$A30 000. Read each card in turn, and determine the maximum salary. Print out the maximum salary and the card number. If there is more than one card corresponding to the maximum salary, just find one of these.

1.3 One of the simplest methods of sorting a file is the *sequential sort*. In this method, the largest file element is found and placed at the top of the list, then the next largest file element is found and placed in second position on the list, and so on. Assume that the N cards in problem 1.2 are all in a file called SAL, and that the file elements are numbered SAL(1), SAL(2), ..., SAL (N). Write a flowchart to sort this file into descending order using a sequential sort. Check your solution with the file $10, $20, $30.

1.4 A new tertiary education entry standard is to be set for students who have completed their final year at school. Students with an IQ of below 90 will be sent to grade III institutions; those with an IQ of between 90 and 115 will be sent to grade II institutions, except that those with an average mark of 80 or more may enter grade I institutions. Those with IQ's in the range 115–165 will attend grade I institutions, except that if their average mark is less than 65 they will attend grade II institutions. Finally, those with an IQ $\geqslant$ 165 will attend Egghead College. Draw up a decision table implementing this policy.

1.5 Repeat problem 1.2, this time drawing a decision table rather than a flowchart.

1.6 Show how the basic control structures in Section 1.5 can be written as decision tables.

1.7 Solve problem 1.2 in a natural language formulation.

1.8 Solve problem 1.4 in a natural language formulation.

2. The BASIC Language

2.1 Introduction

In the early 1960s it became clear that the normal batch processing mode of computer operation — that is, submitting a deck of cards or a reel of paper tape as input and waiting hours or days for output — had severe limitations. There was clearly a need for a method of operation whereby the user could be in constant communication with the computer, usually *via* a console typewriter or visual display unit (v.d.u.). This mode of operation is called the **interactive** mode.

The interactive mode has several obvious advantages. For example, it is possible to test immediately several different approaches to solving a particular problem, and then write out the complete program once the best method of solution is found. Alternatively it may be that several courses of action are possible at a given point in a program, with the next step dependent on the results obtained at a particular stage. In the interactive mode the programmer can arrange for these intermediate results to be displayed and can then specify the next stage of the calculation. Yet another example is the use of a computer as a powerful desk calculator for those jobs for which it would be uneconomical to write a program in, say, FORTRAN to be run in the batch processing mode.

The computer can be a particularly useful teaching aid when used in interactive mode. A student can be asked a sequence of questions which depend on his answers to previous questions; hence this approach to teaching is called *programmed learning*.

To implement interactive or **direct-access** processing successfully, two things are necessary. One is a reasonably simple conversational language. This should be an easily learnt language as close to English as possible, bearing no relation to the machine language — a so-called *high-level* language. In this way the use of a medium to large computer can be made available to a number of users, many of whom may be comparatively ignorant about the finer points of programming.

The second requirement is a **time-sharing operating system**. An *operating system* is the key piece of software in a computing system and is a set of programs designed to

21

maximize the overall effectiveness of the system. By *time-sharing* we mean that the operating system is capable of handling a number of users by sharing out the available time and storage facilities amongst the users. We shall not discuss this second requirement further. Rather, we shall devote the remainder of this chapter to a discussion of the BASIC language.

2.2   The Language Description

BASIC is a very simple high-level scientific programming language that has become the most popular conversational language. The name is an acronym for Beginners' All Purpose Symbolic Instruction Code. The language was developed at Dartmouth College Hanover, New Hampshire, U.S.A., in 1965 by J. G. Kemeny, T. E. Kurtz, and collabora tors, and was originally designed for the G.E. 225 computer. It was designed to be easily learnt and yet to possess such features that allow students to make the transition to more powerful languages such as FORTRAN and ALGOL with minimal difficulty.

The language has been highly successful and widely used, with each manufacturer writing his own version of BASIC. Unfortunately this proliferation, in the absence of any accepted standard, means that there is no unique BASIC language. Most manu facturers provide one or more special features to make their own version more attractive, usually with some name such as Extended Basic, Basic Plus, Super Basic, etc. This makes a language description rather difficult, but the following description of BASIC will consider the features possessed by *most* versions of BASIC, while certain extensions provided by some manufacturers will be briefly mentioned.

Unfortunately, BASIC is a highly unsuitable language in which to exercise the principles of structured programming, due largely to its very limited set of control structures. However we shall see that the six basic control structures introduced in the previous chapter can be written in BASIC using existing language features. If BASIC programs are designed in a top-down manner and then written using these control structures, readable and maintainable programs should result. Since many BASIC programs tend to be short jobs, used only once, it is perhaps not warranted to write especially well-designed, modular programs in these cases. However, if a longer program that is to be regularly used is written in BASIC it is just as important to write a well-styled program in BASIC as in any other language.

2.3   The Language Specification

The following language description is a summary of the language features common to most versions of BASIC.

BASIC Program Input

Input is terminal oriented, but may be presented on cards. There is considerable freedom in statement layout, the only restriction being that there should be one statement per line (or card). There is usually no provision for continuing a statement beyond one line (card). Blanks are ignored within a line – except within a literal string – which gives considerable freedom in layout. Thus 10LETX=5.0 is identical to 10 LET X = 5.0. The latter form is of course vastly preferable being considerably easier to read.

In the remainder of this chapter we will assume input/output takes place on a terminal. The reader can make the necessary changes for card input/output if necessary.

Character Set

Three kinds of characters may be used, as follows.

1)   Alphabetic characters: the 26 upper case letters of the Roman alphabet, A–Z.
2)   Numeric characters: the 10 digits, 0–9.
3)   Special characters: the number of special characters varies from system to system. A common set is the following 12 characters: + − * /) (= . , ' $ (or £) blank. Depending on the keyboard facilities a number of other characters are sometimes available, including the following: $> \geqslant \leqslant < \neq \uparrow$ ".

Constants

No distinction between real and integer constants is usually made. A constant may be written as a signed or unsigned sequence of decimal digits with or without a decimal point. Such a constant may also be followed by an exponent symbol followed by an optionally signed one or two digit integer. Internal commas are forbidden, and in the absence of a sign, the number is assumed to be positive. No scope for complex constants, extended precision constants or logical constants exists. The maximum number of digits allowed in a constant varies from system to system, but tends to be between 7 and 15 digits. Some examples of legal constants are: 7.634, −7.0, −7, 1964.2895, −3.14159265, 123457, −27E−4 (which means $−2.7 \times 10^{-3}$), −27.0E−04 (which means $−2.7 \times 10^{-3}$), 0.163E21 (which means 1.63×10^{20}). The maximum and minimum size of a constant also varies from system to system, and is in the range $10^{-38}–10^{38}$ with the PDP11 Basic Plus system and $10^{-76}–10^{76}$ with the ICL 1900 series system.

Strings

String manipulation is a prominent feature of BASIC. A *string* is a sequence of characters forming a single data value. Strings are usually enclosed in quotation marks (sometimes called *closed strings*), though some systems allow *open strings,* which are strings without quotation marks. Spaces within a string *are* significant. The length of a string is fixed on some systems to at most 15 characters, while other systems allow an entire line or card. Strings usually cannot contain the string delimiter symbol (single or double apostrophe). Some examples of strings are:

> "TODAY I AM 21"
> "I AM A STRING"
> "RING-A-DING-DING, I AM A STRING"

This last example, containing more than 15 characters, is illegal on many systems.

Variables

BASIC **variables** or **identifier names** may comprise one or two characters. The first character must be alphabetic, the second must be numeric. Examples include A, A0, A˙ B, X, Y1, Z5. Thus there are only 26 x 11 = 286 possible variable names. Many system̊ automatically initialize all 286 possible variable names to zero at the start of each run, though it is unwise to rely on this. String variables are denoted by a single letter followed by either a dollar sign or a pound sign. Where both are available, they reference the same string; that is, T$ and T£ are treated as the same symbol. Clearly, there are only 26 possible string variable names, A$–Z$. (Some versions of BASIC allow string variable names to be *any* valid variable name followed by the symbol $, thus allowing 286 string variable names also.) Both simple-array variables and string-array variables exist in BASIC. Simple-array variables may have one or two subscripts, but not more. The variable name must comprise a *single* alphabetic character. Subscripts are separated by commas and enclosed in parentheses. Examples include: A(5), B(10,10), O(5,3). Most systems also allow zero subscripts, so that A(0,0), B(0,5), C(0) would be legal on these systems, though negative subscripts are illegal. Subscripts may also be any valid arithmetic expression, which is rounded to the *nearest integer,* if it is not already an integer.[†] Thus A(K), B(K+2−Q1, B7*17.328), C(X**Y1−2.5) are valid simple array variable elements. String-array variables are valid string names with only one subscript, which specifies a particular string. For example, B$(3) specifies the 4th string (number̊ from 0) in the array B$. Most systems place some limit on array size, a typical limit being 5000 words. The *scope* of a variable name is the entire program, including subroutines.

Arithmetic Operators

Non-string constants and variables may be combined by the usual five arithmetic opera‑ tions of addition (+), subtraction (−), multiplication (*), division (/), and exponentia‑ tion (** or ↑ or both). Some examples are shown below: A+B1+C(5,4), I+J, B9−4.632+C(5,1), B3*Z2, A(5,1)*B(4.2)*9.731*3, A/B(2,7)/C, I1/I2, I1**I(3), I↑B(4.2).

Expressions may be enclosed within parentheses to prevent confusion with sequence̊ of operators and operands. Expressions within parentheses are evaluated first. A **hier‑ archy** of operators exists which usually makes it possible to delete most of the paren‑ theses. This hierarchy is

Hierarchical rank	Operations	Order of operations within hierarchy
3	** or ↑	left to right
2	* and /	left to right
1	+ and −	left to right

[†] This is by no means universal. Many BASIC systems truncate real subscripts towards zero rathe̊ than round them. For this reason it is unwise to utilize this language feature, as it inhibits program portability.

(Unary minus is often not given a hierarchical rank. When it is, it is usually rank 3.) Operations with the highest hierarchical rank are carried out first. Operations of the same hierarchical rank are carried out in the order shown. Thus, $A1/B2*C = (A1*C)/B2$, $A1/A2/A3 = A1/(A2*A3)$. Note that $A1**A2**A3 = (A1**A2)**A3 \neq A1**(A2**A3)$. In this last example it is good programming practice to include the parentheses. Arithmetic operations may not be used between string variables. Division by zero is forbidden, as is raising zero to a negative exponent and raising a negative number to a nonintegral power.

Relational Operators

Most versions of BASIC include six relational operators which may be used between arithmetic expressions and string expressions. It is convenient to think of the value of such an expression as *true* or *false,* even though Boolean variables are not a part of the BASIC language. These six operators are listed below.

Relational operator	Meaning	Example
$<$	is less than	$A < B$
$\leqslant$ or $<=$	is less than or equal to	$A3 \leqslant 40$
$=$	is equal to	$B\$ = C\$$
$\geqslant$ or $>=$	is greater than or equal to	$A(3,5) \geqslant C2$
$>$	is greater than	$A(I+J,K**2) > B(7+X)$
$\neq$ or $<>$	is not equal to	$P2 \neq C3$ or $P2 <> C3$

Some older keyboards do not have these symbols, in which case they are usually replaced, in sequence, by LT, LE, EQ, GE, GT, NE. Similarly, those keyboards without the symbol $\neq$ often use $<>$ for this symbol, as shown in the table above. String variables are usually only compared for equality or inequality, though many versions of BASIC allow other relational operators to be used. In this case corresponding *characters* in a string are compared according to their internal representation in the character code of the particular computer system. Since this is a machine-dependent feature, its use is not recommended.

Arithmetic Matrix Operators

One powerful feature of the BASIC language is its facility to code matrix operations in a simple, direct manner. The three arithmetic operators $+$, $-$, and $*$ may be used between matrices to achieve addition, subtraction, and multiplication of one matrix by another, as shown in the following examples:

$$\text{MAT } A = B+C, \text{ MAT } B = C-D, \text{ MAT } D = G*H$$

where A, B, C, D, G, and H are appropriately sized matrices.

Further, a matrix may be multiplied by a simple arithmetic expression by enclosing the arithmetic expression in parentheses. For example:

$$\text{MAT D} = (\text{NA+B(T)}**2.5)*\text{C}.$$

In this example D and C are matrices. These and other matrix manipulation statements will be discussed further in the section on matrix operators (page 36).

Assignment Statements

There are three types of assignment statements in BASIC: the *arithmetic, string,* and *matrix* assignment statements. The **arithmetic assignment statment** is of the form

$$n \text{ LET v = e}$$

where *n* is an obligatory integral statement number in the range 1–9999 or 1–99999, depending on the system, v is a variable name, and e is an arithmetic expression. Some examples are:

10 LET A = 5
25 LET A(5,3) = 4.0*C(2,2)–B(1)
50 LET A9 = B6

All assignment statements assign to the left-hand variable the value of the right-hand variable or expression. Some versions of BASIC allow multiple-assignment statements, such as 10 LET A = B = C = 5.0.

A **string assignment statement** is

$$n \text{ LET sv = st}$$

where *n* is as above, sv is a string variable name, and st is a closed string or another string variable. For example, 10 LET P$ = 'FRED NURK'. Many BASIC systems allow the deletion of the word LET in an assignment statement.

Discussion of **matrix assignment statements** will be deferred for the moment, though some examples have already been seen in the previous section.

Statement Numbers

All BASIC statements must be prefixed by a statement number, usually in the range 1–99999, though some systems restrict the range to 1–9999, or some other range. Statements are executed in strict numerical order. For this reason it is a good idea to increment statement numbers by 10, allowing for later insertions if necessary. If two (or more) statements bear the same statement number, all but the last statement is ignored. System commands (discussed later) which request operating system action are not considered to be statements, and therefore are not numbered.

Program Annotation

Program annotation or comment is usually provided by the REMARK statement. In fact, any statement beginning with the letters REM is ignored by the compiler. This statement is therefore useful to remind the programmer of the effect of various parts of this program. Some examples:

10 REMARK CALCULATE AVERAGE SALARY
20 REM RIEMANN ZETA FUNCTION, ZETA(K)
30 REMARKABLE THINGS THESE COMPUTERS

Program Control Statements

The natural order of execution of a program is in increasing numerical order of statements, ordered by statement number. Variation in this order is often desirable, and this may be achieved by **program control statements**, which include the following:

GO TO statements

There are two types of GO TO statements: the *unconditional GO TO* and the *computed GO TO*. The unconditional GO TO takes the form k GO TO n where k and n are statement numbers. Example: 35 GO TO 15 transfers control to statement 15. The computed GO TO takes the form m ON ae GO TO $n_1, n_2, \ldots, n_k$ where $m, n_1, n_2, \ldots, n_k$ are statement numbers and **ae** is evaluated and rounded to the largest integer not greater than the value of **ae**. If this integer is p, control is transferred to statement number n_p. If p is outside the range $[1,k]$ an error message is printed and execution ceases. For example, the statement

50 ON X**X−3.6 GO TO 20,30,70,100

transfers control to statement 20,30,70 or 100 depending on whether the integral part of $X^2 - 3.6$ is 1,2,3, or 4.

IF-THEN statements

This statement is written

k IF ea ro eb THEN n

where n and k are statement numbers, **ea** and **eb** are both arithmetic expressions or both string expressions, and **ro** is one of the six relational operators ($<, \leqslant, =, \geqslant, >, \neq$) already discussed. If the Boolean expression **ea ro eb** is *true*, control is transferred to statement n, otherwise normal program flow occurs. Some examples:

10 IF X⩾5 THEN 21
15 IF B**B−4*A*C<0 THEN 37
25 IF A$ = B$ THEN 205

Loop Control

Looping is controlled by a conjunction of a FOR statement and a NEXT statement.
The FOR statement is written

$$k \text{ FOR } lv = \textbf{ea} \text{ TO eb STEP ec}$$
$$k \text{ FOR } lv = \textbf{ea} \text{ TO eb}$$

where k is the statement number, lv is a simple identifier name, and **ea, eb,** and **ec** are
arithmetic expressions. The loop variable lv is initially set to the value of **ea** and statements executed in the usual numerical sequence until the statement m NEXT lv is reached
where m is a statement number, $m > k$, and lv is the *same* simple variable name used
in the FOR statement. The loop is incremented by an amount equal to the value of the
expression **ec** until lv is outside the range [**ea,eb**] whereupon the loop is skipped. Note
that **ec** may be positive or negative and need not be an integer. If positive, then
clearly **eb** should be greater than or equal to **ea**, while if negative **eb** should be less than
or equal to **ea**. If these conditions are not satisfied, the loop is skipped and control is
transferred to the statement numerically following the associated NEXT statement. If
ec has the value zero, program execution ceases and an error message is printed. If **ec**
has the (fixed) value of 1, the second form of FOR statement shown above may be
used. Two examples of a loop to calculate the sum of the first 10 integers are:

(1)   5 S = 0
      10 FOR X = 1 TO 10
      15 S = S+X
      20 NEXT X

(2)   5 S = 0
      10 FOR X = 10 TO 1 STEP −1
      15 S = S+X
      20 NEXT X

The loop variable lv may be used by any statement within the loop and may even be
modified within the range of the loop, though it is poor programming practice to do so.
Loops may be nested to a depth which varies from system to system. The range of two
loops must not intersect, nor can nested loops use the same loop variable. An example
of nested loops to calculate

$$\sum_{n=1}^{5} \sum_{m=1}^{5} \frac{1}{mn}$$

might be

10 S = 0
15 FOR N = 1 TO 5
20 FOR M = 1 TO 5

25 S = S+1/(M*N)
30 NEXT M
35 NEXT N

If the last two statements had their statement numbers interchanged, this would result in an illegal loop structure, since the loop ranges would intersect.

Control Structures

The program control and loop control statements we have just discussed are sufficient to construct the six basic control structures defined in Chapter 1. It should be noted that since there are no statement parentheses in BASIC (no BEGIN and END type statements), we must frequently skip around groups of statements by use of the GO TO statement. That is, the absence of statement parentheses means that their effect has to be simulated by using GO TO statements. This is one of the principal reasons why BASIC and FORTRAN) are poor languages in which to develop the principles of structured programming. We can see how this works in practice by studying the following coding for the basic control structures.

Concatenation
This is achieved by numbering statements to be concatenated in increasing numerical order.

Conditional Clause
The control structure IF C DO S is written in BASIC as follows:

k IF notC THEN n   In this piece of coding, 'notC' means
l statement(s) S   the logical negation of C. If C is
n . . . .   'A=4', then 'notC' is 'A$\neq$4'. k and n
                                 are statement numbers with $k < l < n$.

Alternative Clause
Like the conditional clause, the alternative clause IF C THEN S_1 ELSE S_2 is written using the IF . . . THEN structure, as follows:

k IF notC THEN n_2 This structure shows the use of the
l statement(s) S_1 GO TO in skipping around an
n_1 GO TO n_3 unwanted section of code. 'notC' is
n_2 statement(s) S_2 the logical negation of C, and
n_3 . . . . $k < l < n_1 < n_2 < n_3$.

Choice Clause

The choice clause is represented by the computed GO TO statement.

Thus CASE(i) OF $S_1, S_2, \ldots\ldots, S_N$ becomes

k_1 ON i GO TO $n_1, n_2, \ldots\ldots, n_N$ To implement this structure, the
n_1 statement(s) S_1 various CASEs must be coded by a
k_2 GO TO n_{N+1} simple index i. This requirement
n_2 statement(s) S_2 often involves using many
k_3 GO TO n_{N+1} IF . . . THENs and GO TOs before the
       .   computed GO TO statement is
       .   reached.
k_N GO TO n_{N+1} (See Problem 2.4). Note that
n_N statement(s) S_N $k_1 < n_1 < k_2 < n_2 \ldots k_N < n_N < n$
n_{N+1} . . . .

Repetition and Iteration

Both these control structures can be written in BASIC using an IF . . . THEN statement as shown below:

Repetition: REPEAT S UNTIL C becomes   Iteration: WHILE C DO S becomes
n_1 statement(s) S n_1 IF notC THEN n_4
n_2 IF notC THEN n_1 n_2 statement(s) S
n_3 . . . . n_3 GO TO n_1
                       n_4 . . . .

with $n_1 < n_2 < n_3 < n_4$. The loop structure FOR i=A TO B IN STEPS OF C DO S can be written almost identically in BASIC using the loop control statement FOR in conjunction with a NEXT statement, as follows:

n_1 FOR I=A TO B STEP C
n_2 statement(s) S
n_3 NEXT I
with $n_1 < n_2 < n_3$.

Unconditional Transfer

This structure, which we have already used in writing the above control structures, is written GO TO n, where n is a statement number.

Using the above forms of the basic control structures, it is comparatively straightforward to write a BASIC program from the natural language formulation of the program. We shall now look at the remaining features of the BASIC language.

Stop and End Statements

Both the STOP statement and END statement cause termination of program execution. In addition, the END statement causes termination of compilation. Thus a program may contain several STOP statements, but only one END statement. A program need not contain either statement, since normal termination occurs if an attempt is made to read an empty data file.

BASIC Procedures

There are three types of BASIC procedures: defined functions, inbuilt functions, and subroutines. We will consider each of these in turn.

Defined Functions

These are functions with a single argument defined within a particular program by a single statement of the form DEF FN*l*(*arg*) = **ae**. The function name is FN*l* where *l* is any letter. There are thus at most 26 possible defined functions within any one program. The argument *arg* is a dummy argument, and is any *single letter* variable name. Being a dummy argument, it does not refer to any other variable of the same name, but is local to the defining function. **ae** is any arithmetic expression, usually containing the dummy argument. It may contain other variables, but if so these must have values assigned prior to a function call. Calls are simply made by treating the function as any other variable. The function may be defined anywhere within the program, either before or after the first reference to its use (though notice the restriction given in the sentence before last). Some examples are:

$$10 \text{ DEF FNA(X)} = 1.0/X**2$$
$$20 \text{ DEF FNB(A)} = A**2+FNA(A)$$
$$30 \text{ DEF FNC(P)} = A**2-B*C+P$$

Recursive function definitions are forbidden, so that

$$30 \text{ DEF FNF(N)} = N*FNF(N-1)$$

is illegal. Examples of a function call include the following:

$$105 \text{ LET X} = Y+FNA(5)$$
$$115 \text{ LET B} = FNC(X)-FNC(Y)$$

Inbuilt Functions

There are 11 inbuilt functions available with most BASIC systems, though many manufacturers provide more than this number. The 11 standard functions are shown in tabular form below:

BASIC Name	Function definition	Examples
ABS	Absolute value, X	10 LET P = ABS(X) 20 LET P = ABS(B*B−4*A*C)

SGN	Gives +1 if $X > 0$
	    0 if $X = 0$
	  −1 if $X < 0$
		10 LET Q = SGN(Y)
		20 LET R = SGN(COS(B*B−4*A*C))

EXP	Exponential e^X	10 LET Q = EXP(X)
SIN	Trigonometric sine SIN(X), X in radians	10 LET R = SIN(Y**2)
COS	Trigonometric cosine COS(X), X in radians	1101 LET P = COS(Y+Z)
TAN	Trigonometric tangent TAN(X), X in radians	5 LET Q = TAN(A/B)
ATN	Arctangent of X in range $\left[-\dfrac{\pi}{2}, \dfrac{\pi}{2}\right]$, X in radians	73 LET Z5 = ATN(Z5)
LOG	Natural log, $\log_e(X)$	73 LET P9 = LOG(P8)
SQR	Positive square root of X	420 LET Z1 = SQR(B*B−4*A*C)
INT	The largest integer $\leqslant X$ Thus INT(−7.1) = −8, INT(7.1) = 7	53 LET A = INT(B)
RND	Gives a uniformly distributed random number in the range (0,1)	73 LET C = SQR(RND(X))

Subroutines

If a function is sufficiently complex that it cannot be defined by a single statement, or if it is desired to break up a program into a number of parts, it is possible to write one or more **subroutines** to achieve this. In fact, this is the usual way programs are modularized at the coding level, rather than at the planning level.

The subroutine structure corresponds to what would, in FORTRAN, be called an 'open-coded' subroutine with transfer to, and return from, the subroutine being achieved by GO TO statements, the subroutine being embedded in the body of the main program. Control is transferred to the subroutine by a statement of the form n GOSUB k where n and k are statement numbers, k being the statement number of the first statement of the subroutine. Control is transferred to the statement whose statement number immediately follows the subroutine call statement upon return from the subroutine. If a subroutine calls another subroutine the return addresses are placed in a stack and returns take place on a last-in-first-out basis. At the end of a subroutine a RETURN statement must be used, as this provides the only means of exiting from a subroutine. Subroutines are usually written immediately following a STOP statement or GO TO statement in the main program. This prevents accidental entry to the subroutine. As a trivial but illustrative example, assume we wish to incorporate a subroutine to calculate *n*!

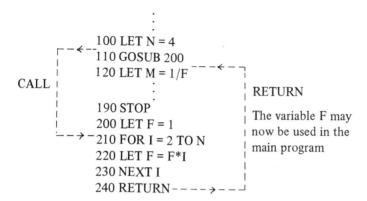

Input, Output, and Data Preparation

Before discussing input instructions, it is necessary to consider how data is provided. This is done by a DATA statement, which takes the form *k* DATA *list* where *k* is a statement number and *list* is a non-empty list of numeric or string constants, separated by commas. If more than one data statement is provided, data is read in the order of increasing statement numbers. If all data has been read, an attempt to read more data causes natural termination of the program.

Data is input by a READ statement, which takes the form *k* READ *list* where *k* is a statement number and *list* is a non-empty list of arithmetic or string variable names. The variable names are placed in 1—1 correspondence with the items in the next executable DATA statement. Data statements may appear anywhere in the program, and are taken in order of their statement number.

```
        :
100 READ A, B
200 READ C$,C
        :
```
cont.

1100 DATA 63.4925,11

⋮

1500 DATA "STRINGS ARE THIN", 3.14159265

This causes A and B to take values 63.4925 and 11 respectively, while the string variable C\$ takes the value "STRINGS ARE THIN", and D takes the value 3.14159265 Once a DATA item is read, it is dropped from the data stack and is no longer accessible. If it is required to access a data list more than once, the RESTORE statement re-establishes the data stack to its condition at the start of execution. Thus, in the followiℸ program segment,

⋮

10 READ A,B
15 RESTORE
20 READ C,D

⋮

100 DATA 3,5

A and C would take the value 3, while B and D would take the value 5. Arrays may be read in by combining a READ statement with a loop, as in the following example:

10 FOR I = 0 TO 10
15 FOR J = 0 TO 20
20 READ X(I,J)
25 NEXT J
30 NEXT I

This reads in 231 elements of array X row by row. Further array input/output statements are discussed in the section on matrices.

Output is achieved by a PRINT statement, which takes the form k PRINT *list* where k is a statement number and *list* is a (possibly empty) list of data items. If the list is empty, a blank line is printed. A data item is any arithmetic expression, any string variable or any closed string.

The value of an arithmetic expression is printed to an accuracy that varies from system to system. Six or seven significant figures are typical. Numbers are printed as integers or floating point numbers unless they are too large or too small to be so represented. In that case they are printed with an exponent part, as previously discussed.

Closed strings are printed in the order given in the PRINT list, but with the quotation marks deleted. The value of string variables is printed similarly. Some examples:

```
90 LET A$ = "THIS IS E"
100 PRINT A,B
200 PRINT
300 PRINT "HERE IS PI",P,A$,E
                 :
                 :
```

Output:

69.2345 1.732E−4

Blank line →

HERE IS PI 3.14159 THIS IS E 2.71828

Most systems provide for more control of the printed output than that given by the PRINT statement alone, but the details of these provisions varies so widely from system to system that no attempt will be made to describe the various methods used, except to mention the tabulating function, TAB, which *is* common to most systems. This function may be included in a PRINT list and takes the form TAB(ae) where **ae** is an arithmetic expression. The largest integer less than or equal to the value of this expression is evaluated *modulo* 75, and the print head moves the appropriate number of spaces before printing the following item in the print list. A print statement using the TAB function might be:

170 PRINT "X="; TAB(10);X; "Y="; TAB(2K−3); Y

Then if K = 55, [2K−3] mod 75 = 32 and the output would appear as

print position 10   print position 32

↓   ↓

X = 6.48921   Y = 19.3417

Notice that semi-colons were used in the above example between items in the print list. This is not mandatory (and indeed, in some systems it is forbidden), but usually produces more closely spaced and better looking output. For further details of formatting of output and of the distinction between commas and semi-colons in a PRINT list, reference should be made to the User's Manual appropriate to your installation.

The final input/output command is the run time input statement, which is used to provide data during the execution phase of a program. This statement is written k INPUT *list* where k is a statement number and *list* is a list of variables to be input. On execution, an 'invitation to type' symbol appears, and the user types a list of data items, separated by commas, in 1−1 correspondence with the list items. It is good programming practice to precede an INPUT statement by a PRINT statement to print out a heading identifying the information to be read in. For example,

>   100 PRINT "TYPE RADIUS OF CIRCLE"
>   110 INPUT R   .
>   .
>   .

Produces the output

>   TYPE RADIUS OF CIRCLE

whereupon the user may type   5.000

If the user-provided list of data items does not agree with the INPUT list, an error message is printed, permitting the user to rectify the error.

Array Statements

As mentioned previously, BASIC arrays may be one- or two-dimensional, and may contain numerical or string information. String arrays, however, may only be one-dimensional. If all array dimensions do not exceed 10, storage allocation is automatic. If any array dimension does exceed 10, storage allocation for that array must be set by a DIM statement, specifying the upper bound on that array. The position of the DIM statement within a program is immaterial. In particular, it need not precede the first array reference. The form of the statement is

$$m \text{ DIM } v_1(n_1), v_2(n_2), v_3(n_3), \ldots, v_N(n_N)$$

where m is a statement number, v_i ($i = 1, N$) are valid variable names and each n_i ($i = 1, N$) is a list of one or two positive integers (only one for a string array), with the list elements separated by a comma. These elements specify the upper bounds on array size. In BASIC arrays are assumed to start at zero, so that a DIM statement of the form

>   115 DIM A(12,12),B(10),A$(5)

would reserve 169 words for array A, 11 words for array B, and six string lengths (often restricted to 15 characters) for the string array A$. An array not listed in a DIM statement is assumed to have a dimension of 10. If the true dimension is less than 10, it saves storage space to give the array size in a DIM statement.

Matrix Manipulation

One of the most powerful features of the BASIC language is the provision of a variety of matrix operations. Matrices are set up in two-dimensional arrays, and those elements with a zero subscript are ignored. That is, in the matrix A(10,10), only those 100 array elements A(1,1) ... A(10,10) are considered to be matrix elements.

The operations of matrix addition, subtraction, and multiplication have already been discussed. For matrix addition and subtraction the dimensions of the two matrices being added or subtracted must of course be the same. For matrix multiplication such as is achieved by the statement 10 MAT X = Y*Z, the dimensions of matrices X, Y, and Z must be in accordance with the rules of matrix multiplication. That is, if X is dimensioned X(m,k), then Y and Z must have dimensions Y(m,n) and Z(n,k).

To read or write a matrix, the MAT READ or MAT PRINT statement may be used. These take the following form:

$$k \text{ MAT READ } v_1(n_1), v_2(n_2), \ldots v_N(n_N)$$
$$l \text{ MAT PRINT } v_1, v_2, \ldots, v_N$$

where k and l are statement numbers, v_i ($i=1,N$) are valid array identifier names and each n_i ($i=1,N$) is an (optional) pair of subscripts that re-dimension the array concerned. If these subscripts are not given, the previously set array dimensions are assumed. The matrix elements are then read row-wise from one or more data statements. On output, three blank lines are printed, then the matrix is printed row by row, with a blank line between rows. Printing starts at the beginning of a line for each new row. An example follows:

```
10 MAT READ A(2,3)
20 MAT PRINT A
          .
          .
100 DATA 1,2,3,4,5,6
```

would read in the matrix A = $\left(\begin{smallmatrix} 1 & 2 & 3 \\ 4 & 5 & 6 \end{smallmatrix}\right)$ and print it out as

```
                    ← 3 blank lines
      1   2   3
                    ← 1 blank line
      4   5   6
```

A matrix may be transposed; that is, matrix element a_{ij} replaced by matrix element a_{ji}, by the following instruction

$$k \text{ MAT } ma = \text{TRN}(mb)$$

where k is a statement number and ma and mb are two appropriately dimensioned different matrices. If ma has dimensions $ma(p,q)$ then mb must be dimensioned $mb(q,p)$.

Transposing a square matrix A by writing MAT A = TRN(A) is forbidden.

The inverse of a square matrix is obtained by writing

$$k \text{ MAT } ma = \text{INV}(mb)$$

with k a statement number and ma and mb two different matrix array names.

A matrix A or a sub-matrix of A, may be initialized by setting all elements to zero or one by the statements

$$k \text{ MAT } ma = \text{ZER}(n)$$
$$l \text{ MAT } mb = \text{CON}(m)$$

where k and l are statement numbers, ma and mb are valid array names and n and m each represent an optional pair of integers that specify new dimensions for the matrix. In the absence of these new dimensions, the previously set dimensions are assumed.

Finally, a matrix identity statement exists which permits the definition of an identity matrix of a specified size (the diagonal has elements = 1, all other elements are zero). The matrix must, of course, be square. The form of this statement is

$$k \text{ MAT } ma = \text{IDN}(n)$$

with k, ma, and n as defined above.

Commands

BASIC commands, by which the user communicates with the BASIC system, are among the most system-dependent feature of the language. The commands discussed here are the most common set, and apply to the original Dartmouth/GE system, ICL Basic, and in large part to PDP 11 Basic Plus, Data General Basic, and Hewlett Packard Basic, amongst others. Operations take place within two files. The first is the user's file, where programs may be saved; the second is the *current file* or *work file* where the program in current use is situated.

At the start of a session, the user is asked

NEW OR OLD?

(In PDP 11 BASIC PLUS, this request is simply replaced by the remark READY.) In reply to either NEW OR OLD? or READY, if the program has previously been saved in the user's file, he responds with the command OLD *name* where *name* is the previously named program under which the program was filed in the user's file. This command retrieves the program and places it in the work file where computation can commence upon the word RUN being typed. If the program has not previously been filed, the user types NEW *name* where *name* is any name the user chooses to identify the program. This is the name under which the program will be filed in the user's file, if desired.

In order to keep a program, the user types SAVE at the end of the program. This causes a permanent copy of the work file contents to be saved in the user's file. It can be later retrieved by the OLD command, followed by the program name, as previously discussed.

The RUN command is typed after a copy of the program is placed in a work file, either by typing OLD *name,* or by typing NEW *name* followed by the program, typed in one line at a time. If the SAVE command is used, it must precede the RUN command. The program name of the SAVEd program must, of course, be unique within the user's file.

To remove a program from the user's file, the instruction UNSAVE *name* is used, where *name* is the program name. To update a program, the program is called into the work file, updated, UNSAVEd, thus destroying the old copy in the user's file, then SAVEd if required, as shown in the following example:

> OLD PROG
> 85 LET X = 3
> UNSAVE PROG
> SAVE

To list the current work file contents, the user simply types LIST. The whole of the work file contents are then listed. To list only part of the work file, the user may specify the statement numbers of the desired statements, or range of statements, as shown in the following example:

> LIST 10,20,30–50,200 TO 500

This prints statements 10 and 20 and all statements in the range 30–50 and 200 to 500.

The DELETE command can be used to remove one or more lines from the current work file. An example is:

> DELETE 10,20,30–50,200 TO 500

which removes all those statement numbers listed following the DELETE command. Effective deletion is also achieved by typing a statement with the same number as the previous statement. This deletes the previous statement.

To change the name of a program, it may be placed in the work file and the instruction RENAME *name* is typed, where *name* is the new name. Notice that in this way we can update a program and keep both the old and new versions, as in the example below:

> OLD PROG 1
> 10 LET X = X+1
> 20 LET Y = Y+1
> 30 LET Z = X+Y−1
> RENAME PROG 2
> SAVE

Thus PROG 1 and PROG 2 are both in the user's file, with PROG 2 having statements 10, 20, and 30 changed from PROG 1.

To terminate a BASIC session and return the user to operating system control, the user types BYE.

Note that all commands are unnumbered, while all statements are numbered.

Most BASIC systems provide a number of other commands in addition to the above. The user should consult the appropriate manual at his installation for details.

Examples

In this section we list some programs and program segments which illustrate much of the foregoing language definition.

(1)   Read 12 numbers into an array and calculate their mean.

```
NEW MEAN
10 DIM A(12)
12 LET S = 0
15 FOR I = 1 TO 12
20 READ A(I)
25 LET S = S+A(I)
30 NEXT I
35 PRINT "MEAN=", S/12
40 DATA 3,4,5,9,14.6,6.7,8.2,9.1,17,11,1,24.8
              .
              .
              .
99 END
SAVE
RUN
```

This program produces the output:

$$MEAN = 9.45$$

It will be SAVEd under the name MEAN, and may be restored to the work file at a later date by typing OLD MEAN.

(2)   A program segment that is part of a program to solve quadratic equations might read:

```
10 LET A = 1
20 LET B = 4
25 DEF FNA(C) = B*B-4*A*C
30 FOR C = 1 TO 5 STEP 2          (cont.)
```

```
32 PRINT
35 IF FNA(C)<0 THEN 55
40 LET S1 = SQR(FNA(C))
45 PRINT "A=",A,"B=",B,"C=","ROOT 1=", (−B−S1)/2/A,
   "ROOT 2=", (−B+S1)/2/A
50 GO TO 60
55 PRINT "A=",A,"B=",B,"C=",C,"COMPLEX SOLUTIONS"
60 NEXT C
        ⋮
RUN
```

This segment produces the following output:

A = 1	B = 4	C = 1	ROOT 1 = −3.73205	ROOT 2 = −0.267949
A = 1	B = 4	C = 3	ROOT 1 = −3	ROOT 2 = −1
A = 1	B = 4	C = 5	COMPLEX SOLUTIONS	

(3) The Newton-Raphson method for obtaining a numerical solution to an equation of the form $f(x) = 0$ states that the sequence

$$x_{n+1} = x_n - \frac{f(x_n)}{f'(x_n)}$$

converges to a root under certain conditions. We wish to write a BASIC program to implement this scheme, with f and f' given by a subroutine, and the starting value x_0 to be read in. Sufficient accuracy is obtained if

$$\left|\frac{x_{n+1} - x_n}{x_{n+1}}\right| < 10^{-5},$$

and if the number of iterations exceeds 200, the program should halt and print out an error message. Given data: $f(x) = e^x - 2x - 21$, so that $f'(x) = e^x - 2$ and $x_0 = 3$.

```
NEW NEWTON
5 REM SOLUTION OF EXP(X)−2*X−21 = 0 BY NEWTON'S METHOD
10 READ X
12 DATA 3
15 FOR I = 1 TO 200
20 GOSUB 100
25 LET X1 = X−F/F1
```

(cont.)

```
30  LET T = ABS((X1-X)/X1)
35  IF T<1E-5 THEN 65
40  X = X1
45  NEXT I
50  PRINT
55  PRINT "NO CONVERGENCE AFTER 200 ITERATIONS"
60  STOP
65  PRINT
70  PRINT "SOLUTION=",X1
75  STOP
100 LET E1 = EXP(X)
110 LET F = E1-2*X-21
120 LET F1 = E1-2
130 RETURN
RUN
```

This produces the output:

$$\text{SOLUTION} = 3.31921$$

Problems

2.1 Write a BASIC program to calculate the greatest common divisor using Euclid's algorithm (defined in Chapter 1). Use the following test data (589,−171) and (5648,8657432) for which the answers are 19 and 8 respectively.

2.2 Write a BASIC program to calculate and print out the first 100 prime numbers, given that the first two primes are 2 and 3.

2.3 Simpson's rule for the numerical approximation to

$$\int_a^b f(x)\,dx$$

is

$$\frac{h}{3}[f(a) + f(b) + 4\{f(a + h) + f(a + 3h) + \ldots + f(b - h)\} + 2\{f(a + 2h) + f(a + 4h) + \ldots + f(b - 2h)\}]$$

where $h = (b - a)/2n$, and $2n$ is the number of strips into which the x axis is divided. Write a BASIC program to evaluate

$$\text{erf}(1) = \frac{2}{\sqrt{\pi}} \int_0^1 e^{-t^2}\,dt$$

using Simpson's rule, with $n = 50$. (Answer: erf(1) = 0.842700793.)

2.4 Write a BASIC program to solve the quadratic equation $ax^2 + bx + c = 0$. Your program should handle the case of complex roots by printing out the real and imaginary parts of the roots. Test your program with the following sets of coefficients.

a	b	c
8.0	−22.0	15.0
0.0	4.0	−5.0
10^{-8}	10^{-4}	10^{-8}
8.0	4.0	0.5
3.0	1.0	2.0

3. The FORTRAN Language

3.1 Introduction

FORTRAN is the oldest and still the most widely used high-level programming language for problems of a scientific nature. The first published version appeared in 1954, issued by the IBM Applied Science Division. FORTRAN — a contraction of FORmula TRANslating system — was developed by a working committee headed by J. W. Backus for implementation on the IBM 704 computer.

An expanded and revised version of FORTRAN, known as FORTRAN II, was issued in 1958. This too was designed specifically for the IBM 704. FORTRAN systems for other IBM computers were released after this, but it was not until 1961 that another manufacturer (UNIVAC) released a FORTRAN compiler, and it was only in 1963 that most manufacturers were releasing FORTRAN compilers for their machines.

Bearing in mind the growth in popularity of the language, it is not at all surprising that almost every manufacturer released slightly different versions of FORTRAN. Indeed, different versions of FORTRAN were implemented on different computers of the same manufacture. It became clear in the early 1960s that some standardization of the language was necessary, and in 1962 the ASA X3.4.3 Committee was formed to define a standard. Their report was published in 1966 and contains two standards: one known as American Standard FORTRAN (which is roughly equivalent to FORTRAN IV), and Basic FORTRAN, which is a proper subset of ASA FORTRAN and is equivalent in scope to FORTRAN II.

At the time of writing, moves are afoot to issue a revised FORTRAN standard. The proposed revisions are largely designed to improve the basic control structures of FORTRAN, thereby making it a better language in which to write well-structured programs.

When the language was developed in the 1950s there was no thought given either to making it machine independent or capable of alphanumeric data handling. Thus the criticisms one often hears about FORTRAN, mainly centered around its deficiencies in these two directions, are largely unjustified. FORTRAN is a good language for what it tries to do. It is easy to learn and to write, and quite fast compilers can be written for it. While there are a number of special-purpose languages which can do most of the things FORTRAN can do faster and more efficiently, none of these combine the generality, popularity, and ease of writing that FORTRAN possesses.

It is a tribute to Backus and his associates that the language they developed in 1954 should be as popular as it is today — albeit fairly extensive development of FORTRAN has taken place since then.

3.2 ASA Fortran Summary

In this section we shall give an informal summary of the language specifications of ASA FORTRAN. We shall also indicate certain language extensions provided by many FORTRAN compilers by mentioning these extensions in parentheses.

FORTRAN Program Input

Input is card-oriented, with the card layout defined in a fairly restrictive manner. Each card is usually assumed to have 80 columns, of which 7–72 inclusive are reserved for a single FORTRAN statement or a part thereof. Blanks anywhere within this field are ignored, except within a Hollerith string. Judicious use of blanks can therefore be employed to increase program readability. If a statement is too long to fit the allowed field of 66 characters, it may be continued onto one or more cards. Up to 19 such continuation cards are allowed, and each continuation is identified by having a character other than zero or blank punched in column 6 of the card. Columns 73–80 are reserved for identification. The contents of these columns are ignored on compilation, except that they are printed out on the output device. Columns 1–5 may be used for an (optional) statement number, in the range 1–99999. If the character **C** is punched in column 1, the card is not translated by the compiler and the card image is printed out. Such a card is known as a *comment* card.

Character set

Three kinds of characters may be used:

(1) Alphabetic characters: the 26 upper-case letters of the Roman alphabet A–Z.
(2) Numeric characters: the ten digits 0–9.
(3) Special characters: there are 11 special characters in ASA FORTRAN;
     + − * /) (= . , $ blank

The set of alphabetic and numeric characters is called alphameric or alphanumeric characters. (A twelfth special character, the single apostrophe ', is often available in extended versions of FORTRAN.)

Constants

(1) *Integers:* These may be preceded by a sign but have no internal commas. Examples: +7, −493, 647912, −1342869. The maximum modulus of such a number is restricted by the word length of each computer.
(2) *Real constants:* These may be written with or without a sign, but must contain either a decimal point or an exponent symbol (the letter E) followed by an (optionally)

signed integer giving the power of ten by which the number is to be multiplied, or both Some examples are: 7.634, −7.0, 1964.2895, −3.14159265, 27.6E+4 (27.6×10^4), −27.E−4(−2.7×10^{-3}), 0.163E 40 (1.6×10^{39}), −0.163E 40 (−1.63×10^{39}), 12E−4 (0.0012).

Double-precision constants are written with the letter D as an exponent indicator, rather than E. Such constants are stored internally with approximately twice as many significant digits as the corresponding single-precision number. The number of significa digits stored in both single and double precision is a machine-dependent feature that varies widely from machine to machine. (Some extended FORTRAN compilers allow for a double-precision constant to be written as a string of digits including a decimal point. It is necessary, however, for the string of digits to be greater than the single-precision word length of the machine. Since this is a machine-dependent feature, it is strongly recommended that all double-precision constants be written in exponent form.) Some examples are: 1.0D0 (1×10^0), −1.638D+04 (−1.638×10^4), 27.6D−43 (2.76×10^{-42}), 4D0 (4.0).

Complex constants are written as a pair of real constants enclosed in parentheses an separated by a comma. The first constant represents the real part and the second const the imaginary part. Some examples are: (1.6,2.3), (1.6E−03,−2.3E4), (1.6,2.3E4) whic represent the following constants: 1.6+2.3i, 0.0016−23000i, 1.6+23000i, respectively.

(3)   *Logical constants:* There are only two logical constants and these are written .TRUE. and .FALSE.

(4)   *Hollerith constants:* These are formed by an integer $n > 0$ preceding the letter H followed by a string of exactly n characters, including blanks. This constant is only defined for use in an argument list in a CALL statement and in a data-initialization statement.

Variables

FORTRAN **variables** or **identifier names** must comprise between one and six alphameri characters, the first of which must be alphabetic.

A predefined type convention exists which assumes that a variable starting with one of the six alphabetic characters, I,J,K,L,M,N is an **integer identifier name**, while those variables starting with the remaining 20 letters of the alphabet are **real identifier names** These conventions can be over-written by a *type declaration statement* − to be discussed later − but it is not usually good programming practice to do so. Variables may also be *subscripted*, and may contain up to three positive non-zero subscripts of integer type. The subscripts are parenthesized, separated by commas, and each must be of one of the following forms: *ic, iv, iv±ic, ic*iv*, ic_1 *iv±ic₂*, where *ic, ic₁*, and *ic₂* are integer constants and *iv* is an integer variable. (These somewhat arbitrary restrictions on the form of a subscript are relaxed by many compilers, which allow combinations like *iv±iv, iv*iv±iv*, etc.) The presence of a subscripted variable must be indicated by a DIMENSION statement, a TYPE statement or a COMMON statement, which will be discussed later. *Logical identifier names* follow the same rules as identifier names, but must be declared as logical identifiers by a *type declaration statement.* Some examples of identifier names follow:

(1) Integer identifiers: I, J, IJ1, IJKLMN, M5N6, K1234;
(2) Real identifiers: A, Z, AB12, ABFGOL, BLOB, DRIP, D1567;
(3) Integer subscripted variables: I(1), MN(1,2,3), L(4*I−2,6),
     IJK(3*I−4,4*J−8,3*M−6);
(4) Real subscripted variables: A(1), Q(1,2,3), ANS(4*I−9,6),
     RESULT (3*J−5,4*N−7,3*J−4).

(Many compilers allow more than six letters in a variable name.) The *scope* of a variable name is the entire program segment (main program or subprogram), except that dummy arguments with names duplicating names of the same type can be used in function and subroutine references. External names such as FUNCTION, SUBROUTINE, and COMMON names have scope over the entire program.

Arithmetic Operators

Constants and variables of the same type (real, double-precision, integer or complex) may be combined by the five arithmetic operations of addition, subtraction, multiplication, division, and exponentiation, thereby forming arithmetic expressions, subject to certain restrictions and cautions as follows. Dividing one integer by another gives as the quotient the largest integer arithmetically less than or equal to the true quotient, for example, $7/4 = 1$ and $−7/4 = −1$. A real, double-precision or complex constant or variable may be raised to an exponent of integer type. If a real or double-precision constant or variable is raised to an exponent of real or double-precision type, then the variable or constant must be non-negative. A complex exponent is not permitted and zero raised to a negative or zero power is undefined. Real and double-precision entities may be combined in the same statement, in which case the single-precision entity will be converted to double-precision before expression evaluation. The five operators and examples of their use are shown below.

Operator	Meaning	Examples
+	Addition	A+B+C, I+J, A1234+ZAP, ION+KLOT+J, A+1.4
−	Subtraction	A−B−C, I−J, A1234−ZAP, ION−KLOT−J, A−1.4
*	Multiplication	A*B*C, I*J, A1234*ZAP, ION*KLOT*J, A*1.4
/	Division	A/B/C, I/J, A1234/ZAP, ION/KLOT/J, A/1.4
**	Exponentiation	A**B, I**J, A**2, A**2.5

To prevent confusion or ambiguities arising with sequences of operators and operands, it is possible to enclose expressions in parentheses. Such expressions are evaluated before any other operators. A **hierarchy** of arithmetic operations exists, which often makes it possible to delete most, if not all, of the parentheses in an expression. This hierarchy is

Hierarchical rank	Operations	Order of operations within hierarchy
3	**, unary −	right to left
2	*, /	left to right
1	+, −	left to right

Operations with the highest hierarchical rank are carried out first. Operations of the same hierarchical rank are carried out in the order shown. Thus $A/B*C = (A*C)/B$, while $A/B/C = A/(B*C)$. Note that $A**B**C$ is not defined. Parentheses must be used indicate whether $(A**B)**C$ or $A**(B**C)$ is required.

Relational Operators

ASA FORTRAN, but not Basic FORTRAN, includes six relational operators which are used between arithmetic expressions to produce the values *true* or *false* for logical variables. These operators are listed below:

Relational operator	Meaning	Examples
.LT.	is less than	A.LT.B
.LE.	is less than or equal to	C123.LE.B45
.EQ.	equal to	IK.EQ.7
.GE.	is greater than or equal to	9.GE.MIN
.GT.	is greater than	ANS1.GT.ANS2
.NE.	is not equal to	BLOB.NE.DROP

The arithmetic expressions connected by relational operators must be of the same type (real or integer, but not complex) though single- and double-precision expressions may occur together, in which case the single-precision variable is converted to the same length as the double-precision variable by right-filling with zeros. Great care should be taken when comparing non-integral quantities for equality.

Logical Operators

Three logical operators permitting the logical operations of conjunction, disjunction, and negation are available in FORTRAN. These are given below:

Logical operator	Meaning	Examples
.AND.	Logical conjunction	(A.GT.B).AND.(C.GT.D)
.OR.	Logical disjunction	(I.EQ.J).OR.(J.EQ.K)
.NOT.	Logical negation	.NOT.(A.GT.B)

In this table, A,B,I,J and K are identifier names. The hierarchical order for logical operators is .NOT., .AND., .OR., in that order. For example, the expression L = C.OR.D.AND.A.OR. . NOT.C.AND.E is equivalent to
L = C.OR.(D.AND.A).OR.((.NOT.C).AND.E).

Assignment Statements

There are three types of assignment statements in FORTRAN; the *arithmetic, logical,* and *go to.* The **arithmetic assignment statement** is of the form $v = e$, where v is a variable name and e is an arithmetic expression. Certain type mixing is allowed and the results of the allowed type mixing is shown in the table below:

Type of v \ Type of e	Integer	Real	Double precision	Complex
Integer	Integer	Fix and assign	Fix and assign	Prohibited
Real	Float and assign	Real	DP evaluate and real assign	Prohibited
Double precision	DP float and assign	DP evaluate and assign	Double precision	Prohibited
Complex	Prohibited	Prohibited	Prohibited	Complex

Assignment rules for arithmetic assignment statement $v = e$ in FORTRAN: Note that 'Fix' means to truncate a real number, retaining only the integral part; 'Float' means to convert an integer to a real number; 'DP' means double precision.

The **logical assignment statement** is also of the form $v = e$ where v is a logical variable and e is a logical expression. The logical variable takes the value .TRUE. or .FALSE. as a result of this assignment statement.

The **go to assignment statement** is of the form ASSIGN m TO n where m is a statement name and n is an integer variable name. This statement must be combined with a statement of the form GO TO n, $(l_1, l_2, l_3, \ldots, l_k)$ where n is an integer variable and l_k is a statement label. Example: ASSIGN 10 TO K followed later in the program by GO TO K, (1,10,50,80,200) results in an unconditional transfer to statement 10.

The go to assignment statement is considered redundant by many programmers, since the computed GO TO statement, discussed below, does everything that this assignment can do. On occasion, however, the *go to* assignment statement is more convenient than the computed GO TO statement.

Program Control Statements

The 'natural' order of execution of a program is sequential — that is, the statements are executed in the order in which they are given. In almost all programs some variation in this order is necessary and this is achieved by **program control statements**, some of which we have alluded to earlier. Program control statements include the following:

GO TO statements

There are two types of GO TO statements in FORTRAN. The first achieves unconditional transfers and takes the form GO TO m, where m is a statement number. Exampl
GO TO 100 transfers control to statement number 100.

The second GO TO statement is the computed GO TO, which takes the form GO T
$(n_1, n_2, \ldots, n_k)$, i where the n_i's are statement numbers and i is an integer variable. Th
statement effects a transfer to statement number n_i. For example, GO TO (10,30,100,
ISNUM will transfer control to statement number 10 if ISNUM = 1, to statement 30 if
ISNUM = 2, and so on. Clearly, the statement is undefined if i is outside the range $[1,k$

IF statements

These statements are used to write the iterative, repetitive, conditional, and alternative
clauses, and come in two flavours as follows:

(1)   Arithmetic IF statements take the form IF(e)n_1, n_2, n_3 where e is any arithmetic
expression and n_1, n_2, n_3 are statement numbers of executable statements. Control is
transferred to statement n_1 if the value of $e < 0$, to n_2 if the value of $e = 0$, and to n_3
if the value of $e > 0$. For example, IF(4$-$I)10,20,30 transfers control to statement 10
if I > 4, to statement 20 if I = 4, and to statement 30 if I < 4. The statement numbers
n_1, n_2, and n_3 need not all be different. Thus the arithmetic IF statement can be used
to code two or three way branches. Since the coding of branches in this manner tends
to lead to highly unreadable programs, this statement is falling into disuse. Note that
with all three statement numbers the same, we have a highly inefficient way of writing
GO TO n.
(2)   The logical IF statements take the form IF(**be**)S where **be** is a logical expression
and S is any executable statement. If **be** is *true*, statement S is executed. If **be** is *false*,
statement S is treated as if it were a CONTINUE statement. For example,

$$\text{IF(K.L.T.0) K} = \text{I}-\text{K}$$

replaces K by I $-$ K if K < 0 and has no effect if K $\geqslant 0$.

DO statements

Looping can be controlled by logical IF statements, combined with GO TO statements
but the DO statement is a particularly compact way of setting up a loop and has the
further advantage that most compilers produce a very efficient object code from a DO
loop. The DO statements take the form

$$\text{DO } n \; lv = i, j, k$$

or

$$\text{DO } n \; lv = i, j$$

where *n* is a statement label of a statement which physically follows the DO statement and specifies the *range* of the loop. *lv* is the loop variable and is a non-subscripted integer variable. *i, j,* and *k* are integers or integer variables, all of which are positive, and it is necessary that $j \geq i$. (Most compilers will accept $i > j$, but the interpretation varies.) This statement is equivalent to the control structure FOR $lv = i$ TO j IN STEPS OF k DO, where the range of the loop is all statements between the DO statement and statement number *n* inclusive. For example, with

$$\text{DO } 200 \text{ I} = \text{J},100,5$$

$$\vdots$$

$$200 \text{ CONTINUE}$$

the statements down to statement 200 are excuted with I = J, then this sequence of statements is executed with I = J+5, then I = J+10, then I = J+15, and so on. The last time the statements within the loop are executed is with $95 < I \leq 100$.

The second form of the DO statement

$$\text{DO } n \; lv = i, j$$

is identical to DO n $lv = i, j, 1$.

Certain restrictions must also be observed when using DO loops. It is forbidden to jump into the range of a loop from outside — for the obvious reason that the loop variable is then undefined.

It is also forbidden to redefine the loop variable within the loop, and if any of *i, j, k* are given as variable names, rather than integer constants, these too must not be re-defined within the range of the loop. It should also be noted that after *completion* of a loop, the loop variable is undefined. If a loop is exited before completion the loop variable is then defined however.

The final statement of a DO loop, with statement number *n*, must not be a GO TO statement, arithmetic IF, RETURN, STOP, PAUSE, DO, or a logical IF containing one of these forms.

Finally, DO loops may be **nested**. That is, the range of one or more DO loops may be contained within the range of another DO loop. The range of DO loops must not overlap. For example, the following arrangements are permitted.

(1)	DO 100 I = 1,100,3	(2)	DO 200 N = K,L,5
	DO 200 J = 2,100,4		DO 200 I = J,K,N
	$\vdots$		DO 100 M = 5,10
	$\vdots$		$\vdots$
	200 CONTINUE		100 CONTINUE
	100 CONTINUE		200 CONTINUE

whereas the following arrangement is forbidden, as the loop variables overlap:

$$\text{DO } 100 \text{ I} = 1,300,6$$
$$\text{DO } 200 \text{ J} = 2,\text{K},\text{L}$$
$$\vdots$$

$$100 \text{ CONTINUE}$$
$$200 \text{ CONTINUE}$$

CONTINUE statement

This statement causes continuation of the normal execution sequence. It is in essence a dummy statement, producing no change in any variable. It is good programming practice to end all DO loops on a CONTINUE statement. Since blanks are ignored with a statement, a recommended practice is to indent all statements in the range of a loop. This results in a more readable program.

Control Structures

At this stage of our discussion of the FORTRAN language we have defined all the statements needed to code the basic control structures introduced in Chapter 1. The comments we made about control structures in BASIC largely apply to FORTRAN also. That is, since there are no statement parentheses, GO TO statements must be used to skip around groups of statements. (This is one deficiency receiving attention in the current revision of FORTRAN.)

Concatenation

This is the normal method of execution of sequential statements.

Conditional and Alternative Clauses

The conditional control structure IF C DO S and the alternative control structure IF C THEN S_1 ELSE S_2 can both be written using the logical IF statement as follows:

(1)   If S is a single statement, write IF(C) S for the conditional clause. Otherwise, write

$$\text{IF (notC) GO TO } n$$
$$\text{statement(s) S}$$
$$n \ldots \ldots$$

where (notC) is the logical negation of C.

(2)   The alternative control structure can be written

$$\text{IF (notC) GO TO } n_1$$
$$\text{statement(s) } S_1$$
$$\text{GO TO } n_2$$
$$n_1 \quad \text{statement(s) } S_2$$
$$n_2 \quad \ldots \ldots$$

The Choice Clause

The control structure CASE(i) OF $S_1, S_2, \ldots, S_N$ can be written in FORTRAN using the computed GO TO, as shown below.

$$
\begin{array}{ll}
& \text{GO TO } (n_1, n_2, \ldots \ldots, n_N), i \\
n_1 & \text{statement(s) } S_1 \\
& \text{GO TO } n_{N+1} \\
n_2 & \text{statement(s) } S_2 \\
& \text{GO TO } n_{N+1} \\
& \quad \vdots \\
n_N & \text{statement(s) } S_N \\
n_{N+1} & \ldots \ldots
\end{array}
$$

Note that the various cases must be controlled by a single index i. To achieve this often requires some quite clumsy programming before the computed GO TO statement is reached. This point is also discussed in the previous chapter, where similar remarks were made about this control structure in BASIC.

Repetition and Iteration

The subclass of iterative problems that can be controlled by a simple loop with positive loop increment can be handled by a DO loop in FORTRAN, as defined above. The general iterative and repetitive control structure is written with a logical IF statement. The repetitive clause REPEAT S UNTIL C becomes:

$$
\begin{array}{ll}
n_1 & \text{statement(s) } S \\
& \text{IF (notC) GO TO } n_1
\end{array}
$$

The control structure WHILE C DO S can be written similarly as

$$
\begin{array}{ll}
n_1 & \text{IF (notC) THEN } n_2 \\
& \text{statement(s) } S \\
n_2 & \text{GO TO } n_1 \\
& \ldots \ldots
\end{array}
$$

Unconditional Transfer

This structure, which we have already used in writing the above control structures, is written GO TO n, where n is a statement number.

This completes our discussion of the basic control structures in FORTRAN. In the remainder of this chapter we will study other language features.

FORTRAN Procedures

There are four kinds of FORTRAN procedures: statement functions, inbuilt functions, external functions, and external subroutines. We will discuss these in order.

Statement Functions

These are defined within a particular program by a single statement of the form

$$F(a_1, a_2, \ldots, a_n) = E$$

where F is the function name, $a_1 \ldots a_n$ are dummy arguments specifying the number and type of arguments, and E is an expression. For example,

$$DISCR(A,B,C) = SQRT(B*B-4.0*A*C)$$

defines the function DISCR with arguments A,B, and C to be equal to $(B^2-4.A.C)^{\frac{1}{2}}$. A any point in the program a reference to a statement function may be made by using th function name on the right-hand side of an assignment statement. The arguments may be constants or variables, but if they are variables they must have had values assigned t them before they are used. Examples of calls to the statement function just given are

$$ROOT1 = -5.0+DISCR(1.0,5.0,4.0)$$

and

$$ROOT1 = -Y+DISCR(X,Y,Z)$$

In the second example, the values of X, Y, and Z must have been previously assigned.

Statement functions must precede all executable statements. The expression E may contain only the dummy arguments, which must not be subscripted, constants and variables of the correct type, references to inbuilt functions and references to previousl defined statement functions.

Inbuilt Functions

There are a number of inbuilt functions provided by FORTRAN which are summarizec in the table on pages 55–57.

The two remaining kinds of procedures to be discussed, FUNCTION subprograms and SUBROUTINE subprograms, differ from those already considered in that they are totally separate entities. They are, in fact, *subprograms*, and they can be compiled independently of the main program. Their variable names are independent of those in the main program and those in other subprograms. This has the enormous advantage that a subroutine written by one programmer can be used without modification by another programmer.

Function Subprograms

If a function is sufficiently complex that it cannot be defined by a single statement – and therefore cannot be defined by a statement function – FORTRAN allows a separate subprogram, called a **function subprogram**, to define the function.

The first statement must be of the form

$$t \text{ FUNCTION } f(a_1, a_2, \ldots, a_n)$$

Function definition	FORTRAN name	Argument type	Function type	†In-line (I) or Out-of-line (O)	Example
Absolute value $\lvert a \rvert$	IABS	Integer	Integer	I	JK = IABS(M+N)
	ABS	Real	Real	I	X = ABS(X+50)
	DABS	DP	DP	I	DX = DABS(DX+DY)
	CABS	Complex	Complex	O	CX = CABS(X*Y)
Convert from integer to real (or double precision)	FLOAT	Integer	Real or DP	I	XJ = FLOAT(IX+J)
Convert from real (or double precision) to integer [X]	IFIX	Real	Integer	I	JY = IFIX(X)
Sign transfer $\lvert a \rvert \times$ sign (b)	ISIGN†	Integer	Integer	I	I = ISIGN(IA,IB)
	SIGN	Real	Real	I	X = SIGN(A,B)
	DSIGN	DP	DP	I	DX = DSIGN(X,Y)
Exponential e^X	EXP	Real	Real	O	X = EXP(Z)
	DEXP	DP	DP	O	Z = DEXP(4.6DO)
	CEXP	Complex	Complex	O	CXP = CEXP(A–B)
Natural logarithm $\log_e(X)$	ALOG	Real	Real	O	X = ALOG(Y)
	DLOG	DP	DP	O	Z = DLOG(6.9DO)
	CLOG	Complex	Complex	O	C = CLOG(A*B)
Trigonometric sin: $\sin(X)$	SIN	Real	Real	O	X = SIN(Y)
	DSIN	DP	DP	O	Z = DSIN(X+Y)
	CSIN	Complex	Complex	O	C = CSIN(DR)
Trigonometric cosine: $\cosine(X)$	COS	Real	Real	O	Y = COS(Y)
	DCOS	DP	DP	O	Z = DCOS(X+Y)
	CCOS	Complex	Complex	O	C = CCOS(CX*CY)

Function definition	FORTRAN name	Argument type	Function type	†In-line (I) or Out-of-line (O)	Example
Hyperbolic tangent tanh(X)	TANH	Real	Real	O	X = TANH(Y)
Square root$(X)^{\frac{1}{2}}$	SQRT	Real	Real	O	X = SQRT(Y)
	DSQRT	DP	DP	O	Y = DSQRT(A+B)
	CSQRT	Complex	Complex	O	Z = CSQRT(Z+X)
Arctangent arctan (X)	ATAN	Real	Real	O	X = ATAN(Y)
	DATAN	DP	DP	O	Z = DATAN(CT+BM)
Arctangent (X_1/X_2)	ATAN2	Real	Real	O	XA = ATAN2(Y1,Y2)
	DATAN2	DP	DP	O	DXA = DATAN2(DY1,DY2)
Truncation Largest integer $\leqslant |a|$ with sign of a	INT	Real	Integer	I	I = INT(X)
	AINT	Real	Real	I	AI = AINT(Y)
	IDINT	DP	Integer	I	Z = IDINT(P)
Remaindering $a_1 \pmod{a_2}$	AMOD†	Real	Real	I	X = AMOD(A1,A2)
	MOD	Integer	Integer	I	I = MOD(K,L)
	DMOD	DP	DP	O	DX = DMOD(DA1,DA2)
Find maximum of $\geqslant 2$ elements	AMAX0	Integer	Real	I	X = AMAX0(IJ,K,L,M,N)
	AMAX1	Real	Real	I	X = AMAX1(A,B,C,D)
	MAX0	Integer	Integer	I	I = MAX0(I,K,12)
	MAX1	Real	Integer	I	J = MAX1(A2,C3,5.0)
	DMAX1	DP	DP	I	Z = DMAX1(C3,4.0D1,8)
Find minimum of $\geqslant 2$ elements	AMIN0	Integer	Real	I	X = AMIN0(I,J,K)
	AMIN1	Real	Real	I	X = AMIN1(A,D,C)
	MIN0	Integer	Integer	I	I = MIN0(I,K,L)
	MIN1	Real	Integer	I	J = MIN1(A3,C2,5.0)
	DMIN1	DP	DP	I	Z = DMIN1(C3,4.0D0)

Function definition	FORTRAN name	Argument type	Function type	†In-line (I) or Out-of-line (O)	Example
Positive difference $a_1 - \text{Min}(a_1, a_2)$	DIM IDIM	Real Integer	Real Integer	I I	X = DIM(Z,P) I = IDIM(K,L)
Obtain most significant part of DP argument	SNGL	DP	Real	I	X = SNGL(Y)
Obtain real part of complex argument	REAL	Complex	Real	I	X = REAL(CX)
Obtain imaginary part of complex argument	AIMAG	Complex	Real	I	Y = AIMAG(CX)
Express single precision argument in DP form	DBLE	Real	Double	I	Z = DBLE(3.0*X)
Express two real arguments in complex form Z = X+iY	CMPLX	Real	Complex	I	Z = CMPLX(X,Y)
Obtain conjugate of a complex argument	CONJG	Complex	Complex	I	Z1 = CONJG(Z2)
Common logarithm $= \log_{10}(X)$	ALOG10 DLOG10	Real Double	Real Double	O O	X = ALOG10(Y) DX = DLOG10(DY)

† Note that AMOD, MOD, SIGN, ISIGN, and DSIGN are not defined if the second argument is zero. Those functions that are described as 'In-line' insert the appropriate machine language commands into the object program whenever a function call to an 'In-line' function is encountered. 'Out-of-line' functions, such as EXP, usually require quite a few machine language instructions, so are handled 'Out-of-line'. That is, the appropriate instructions are kept in a separate area of memory from the user's object program and are invoked at each reference. Such functions are usually executed much more slowly than 'In-line' functions.

where f is the function name and $a_1 \ldots a_n$ are the dummy function arguments. t is an (optional) type specification and may be either INTEGER, REAL, COMPLEX, DOUBLE PRECISION, LOGICAL or is empty, in which case the predefined type convention applies to the function name f. The dummy arguments may be simple or array variable or external subprogram names. The subprogram contains the code for the procedure calculating the function and must contain the function name f on the left-hand side of an assignment statement and at least one RETURN statement. The last statement must be an END statement.

A function subprogram is used in the same way as a statement function. It is merely referred to in a main program.

Example
Write a function subprogram to define the step function

$$S(K) = \begin{cases} -1 & K < 0 \\ 0 & K = 0 \\ 1 & K > 0 \end{cases}$$

The function is to be of integer type.

```
FUNCTION ISTEP(K)
ISTEP = 0
IF (K .EQ. 0) RETURN
ISTEP = K/IABS(K)
RETURN
END
```

The RETURN statement causes control to be returned to the next statement immediately following the subprogram invoking statement. A statement such as IV = N+ISTEP(IP) in the main program or any other subprogram is an example of a subprogram calling statement.

Subroutine Subprograms
A subroutine subprogram is usually used rather than a function subprogram when we require more than one value to be calculated, or when we are calculating something that is not really a function; for example, the transpose of a matrix. The first statement of subroutine is

$$\text{SUBROUTINE } S(a_1, a_2, \ldots, a_n)$$

or simple

$$\text{SUBROUTINE } S$$

S is the subroutine name, which must not be used in the body of the subroutine. The ummy arguments $a_1 \ldots a_n$ are simple or array variables or subprogram names. The broutine need have no arguments, or the arguments can be passed by a COMMON atement, to be discussed later.

After the subroutine heading, the required procedure is coded. In a similar fashion the function subprogram, there must be one or more RETURN statements in the ody of the subroutine, and the final statement must be an END statement.

To invoke a subroutine from another program or subprogram, a CALL statement is cessary. This takes the form

$$\text{CALL } S(b_1, b_2, \ldots, b_n)$$

here S is the subroutine name and $b_1 \ldots b_n$ is the argument list which must agree in mber and type with the argument list in the SUBROUTINE statement.

As a trivial example of a subroutine subprogram we assume that the sum, difference, d product of two integers I and J are required. These are provided by the following broutine:

```
SUBROUTINE ISUB(I,J,ISUM,IDIF,IPROD)
ISUM = I+J
IDIF = I-J
IPROD = I*J
RETURN
END
```

his subroutine may be called from the main program or from another subprogram by e following statement:

```
CALL ISUB(K,L,IS,ID,IPR)
```

which case IS, ID, and IPR return respectively the sum, difference, and product of e two integers K and L.

The RETURN statement causes program control to return to the statement imme- ately following the CALL statement.

Subprograms may not call themselves. The significance of this limitation is fully scussed in Chapter 5. (A few advanced FORTRAN compilers do allow such recursive broutine calls, however.)

There exists a special type of subprogram, called a BLOCK DATA subprogram, which discussed later.

put and Output

here are two types of input and output statements, and these are:

() READ and WRITE statements, and

2) Auxiliary input/output statements, namely, REWIND, BACKSPACE, and ENDFILE.

The READ and WRITE statements may be formatted or unformatted. If formatted, they take the form

READ(n,m)   or   READ(n,m) *List*
WRITE(n,m)   WRITE(n,m)*List*

where n is the peripheral device number on which input or output is sought. The device referred to by a particular value of n varies from installation to installation. The number m is the FORMAT statement number. The FORMAT statement specifies the way the input/output is to be read/written from or onto the specified peripheral. A *list* of variables may be given, in which case the numerical value taken by the listed variables will be read/written by the input/output statement.

Unformatted output is often used to write or read onto or from a peripheral such as tape unit, disc drive or drum. The form of read/write statement is then

READ(n) *List*
WRITE(n) *List*

where n and *List* are as defined above. A non-empty *list* is mandatory for unformatted WRITE statements.

A *list* may be a simple list of variables and array names separated by commas; it may be a list enclosed in parentheses, two lists separated by commas; or a DO *implied list*. A DO *implied list* is a list followed by a comma and then followed by a sequence of characters of the form I = M1,M2,M3 or I = M1,M2. The integers or preassigned integer variables M1,M2, and M3 are defined as for a DO loop. The second form above, with M3 omitted, is assumed to be equivalent to the first form with M3 ≡ 1. The **range** of the DO is the set of parenthesized names preceeding it, with elements specified for each value of the loop variable I. Some examples of READ and WRITE statements are:

WRITE (6,100)A
WRITE (6,110)A,B,C,D
READ (3,591)A,B,(C(I),I=1,10)
READ (4,265)((A(I,J),J=1,L),I=1,K),C,B

The first READ statement reads A,B then C(1),C(2),C(3), . . . up to C(10). The second READ statement reads A(1,1),A(1,2), . . .,A(1,L) then A(2,1),A(2,2), . . .,A(2,L), . . . then A(K,1),A(K,2), . . .,A(K,L) then C and finally B.

Auxiliary Input/Output Statements

The REWIND statement takes the form REWIND n, and causes peripheral device number n to be positioned at its starting point.

The BACKSPACE statement is written BACKSPACE n, and causes device number n

o be positioned one record prior to its present setting. If there are no prior records, he statement has no effect.

The ENDFILE statement, written ENDFILE n, causes an end-of-file mark to be ritten on device number n. An end of file mark signifies the end of a file, or string of gical records.

These three statements are usually used to refer to magnetic tape drives, discs or rums.

ormat Statements

f a formatted input/output statement is used, it is necessary to have an associated ORMAT statement which converts and edits information between the internal and xternal representation of the data. A FORMAT statement takes the form:

$$nn \text{ FORMAT } (q_1 t_1 z_1 t_2 z_2 \ldots t_n z_n q_2)$$

where (a) $(q_1 t_1 z_1 t_2 z_2 \ldots t_n z_n q_2)$ is the format specification,
(b) each q is a series of slashes (/) or is empty,
(c) each t is one or more field descriptors,
(d) each z is a field separator (one or more slashes or a comma),
nd (e) $n \geqslant 0$.

n is the FORMAT statement number. There are nine types of field descriptor, which re summarized in the following table.

FORTRAN Field Descriptors

Descriptor	Type of data	Examples
$rFw.d$	Real or complex fixed point	F6.3, e.g. 27.342 1P2F7.3, e.g. 273.420 64.981 $w \geqslant d+2$
$rEw.d$	Real or complex floating point	E12.3, e.g. 0.273E+02 1P2E12.3, e.g. 2.734E+01 6.498E+01 $w \geqslant d+7$
$rGw.d$	Real or complex fixed or floating point	3G12.4, e.g. 12.451 345.6 2.1 3G12.4, e.g. 984.32 0.6342E+18 4.639 This is the universal real field descriptor. d is the number of significant digits.
$rDw.d$	Double precision	D12.3, e.g. 0.273D+02 This is exactly the same as E format, except that it applies to double-precision identifiers.
Iw	Integer	I4, e.g. −683 3I3, e.g. −27 1 −9

FORTRAN Field Descriptors (continued)

Descriptor	Type of data	Examples
rLw	Logical	L4, e.g. T 4L4, e.g. T F F T
rAw	Characters	A4, e.g. FRED 7A4, e.g. THIS IS THE SATURN DATA NO.4
$nHc_1c_2 \ldots c_n$	Hollerith field	9H ANSWERS, e.g. ANSWERS 12H\$17.00 SALES, e.g. \$17.00 SALES
nX	Blanks	4X, e.g. 12X, e.g.

In this list S is an (optional) scale factor designator and takes the form kP where $k = 1,2,3,\ldots$ It can be disastrous when used with F format, since k represents the power of 10 by which the number is multiplied. Another pitfall is that the effect of scale factors continues for the whole format scan unless 'switched off' by an 0P scale factor on a later entry in the same format list. With E or D format, the scale factor allows significant digits to appear on the left of the decimal point, with the correct power of 10 being given in the exponent part.

r is an (optional) repeat counter. Writing r (field descriptor) is equivalent to writing

$$\underbrace{(\text{field descriptor}), (\text{field descriptor}), \ldots, (\text{field descriptor})}_{r \text{ times}}$$

w is the *field width*; that is, the total number of characters in the field. The output is *right adjusted* in the field and the field is then *left filled* with spaces.

d is the number of digits required on the right of the decimal point in writing a real, complex or double-precision number, so that $d \geqslant 0$.

On input, a decimal point in the input field overrides the decimal point specification supplied by the field descriptor. (Many compilers permit Hollerith fields to be specified by enclosing the Hollerith string in apostrophes; for example, 'ANSWERS'. This form is less error-prone than the nH *string* form, since no count of the field length is necessary.)

$c_1 \ldots c_n$ represents a string of n characters, each of which may be any of the 47 characters in the FORTRAN character set.

The effect of a slash is to indicate the end of a record. A string of slashes is interpreted as having blanks between them. With line printer output a string of n slashes produces a carriage return and a skip of n lines. A comma merely indicates the end of a field.

Field descriptors and groups of field descriptors may be parenthesized and then combined with an (optional) repeat counter. For example,

3(F6.3,I4) is equivalent to
F6.3,I4,F6.3,I4,F6.3,I4

With line printer output, the first character is taken to be a carriage control character. It should be one of the following:

Character	Meaning
Blank	Take a new line
0	Take two new lines
1	Advance to head of a new page
+	Do not take a new line

The last carriage control character listed above is usually only used to achieve special effects, such as printing of θ by superimposing a zero (0) and a bar ($\bar{9}$). The operation of the format statement is such that successive items in the read/write list are transmitted according to the successive field descriptors in the corresponding format statement. If the list items exhaust the number of field descriptors, the format list is rescanned from the first field descriptor in the last set of parentheses in the format statement.

Examples

$$100 \text{ FORMAT}(1H,2F7.2,I4,1PE12.3,2L2)$$

may produce the output

$$1.43\ -16.94\ \ 29\ \ 9.463E-03\ T\ F$$

on a new line. The format statement

$$200 \text{ FORMAT } (28H1OUR\ RESULTS\ ARE\ GIVEN\ BELOW/)$$

would produce the heading OUR RESULTS ARE GIVEN BELOW at the head of a new page, followed by a blank line.

Storage Allocation Statements

There are a number of statements which allow the programmer limited control of the allocation of storage within a program. These are the DIMENSION, COMMON, EQUIVALENCE, and BLOCK DATA statements.

DIMENSION statements

This is a statement of the form

$$\text{DIMENSION } v_1(n_1), v_2(n_2), \ldots, v_n(n_n)$$

where $v_1, v_2, \ldots, v_n$ are array variable names and $n_1, n_2, \ldots, n_n$ are lists of one, two or three positive, non-zero integers separated by commas. This statement sets aside storage space for the specified arrays, identified by their variable names. For example,

$$\text{DIMENSION A(10),B(10,10),ID(3,3,4)}$$

sets aside ten locations for the one dimensional real array A, 100 locations for the two dimensional real array B, and 36 locations for the three dimensional integer array IC. Array elements are referred to by giving their variable name and array index: e.g. A(4), B(5,5), ID(1,1,2) refer to specific array elements.

All subscripted variables must be listed in a DIMENSION statement and the DIMENSION statement(s) must precede all executable statements within a program segment.

COMMON statements

This is a statement of the form

$$\text{COMMON}/c_1/v_1/ \ldots /c_n/v_n$$

where $v_1,v_2, \ldots,v_n$ are lists of variable names and each c_i is a symbolic name, or is empty. If empty, the two slashes are optional. The c_i's are *block names* and these names are independent of any variables which may have the same name. The variable(s) v_i following block name c_i are said to be in common block c_i. Variables within a list v_i are separated by commas. In the absence of a block name, the variables are in *blank common* or *unlabelled common.*

Each COMMON statement sets aside a region of memory that is accessible to all parts of the program including subprograms, which contain a COMMON statement with the appropriate label for that COMMON region.

COMMON statements are usually used to pass information to different subprograms in an efficient manner and to make the most economical use of available storage, since the same region of storage may contain different variable values at different stages of program execution.

EQUIVALENCE statements

This is a statement of the form

$$\text{EQUIVALENCE } (l_1), (l_2), \ldots , (l_n)$$

where l_i is a *list* of at least two variable names or array elements. The statement permits the sharing of storage by two or more entities. For example, the statements

$$\text{DIMENSION VEC(16),XMAT(4,4)}$$
$$\text{EQUIVALENCE(VEC(1),XMAT(1,1))}$$

cause 16 storage locations to be set aside, which may be referred to individually as XMAT(I,J) or VEC(K), where $K = I + 4*(J - 1)$. When two variables share storage because of an EQUIVALENCE statement, their names must not both appear in COMMON statements in the same program unit. It is important to clearly distinguish between EQUIVALENCE and COMMON statements. EQUIVALENCE statements allow two or more variables to share the same storage locations, while a COMMON statement makes the named variables accessible to all parts of the program which reference that COMMON block.

These three statements, DIMENSION, EQUIVALENCE, and COMMON, must precede any statement functions which must in turn precede all executable statements in a program or subprogram.

BLOCK DATA and DATA INITIALIZATION statements

The BLOCK DATA statement is the first statement in a special subprogram called a BLOCK DATA subprogram. The function of this subprogram is to initialize elements of labelled COMMON blocks. No executable statements are allowed in BLOCK DATA subprograms. For example, a BLOCK DATA subprogram to initialize the previously declared COMMON block, CONST, which is to contain the value of π and e, may be written:

```
            BLOCK DATA
            COMMON/CONST/PI,E
            DATA PI,E/3.1415926536,2.7182818285/
            END
```

In this subprogram we have used, for the first time, a **data initialization statement**, which takes the form

$$\text{DATA } l_1/c_1/,l_2/c_2/, \ldots ,l_n/c_n/$$

where l_i is a *list* of variable names and c_i is a one-to-one corresponding *list* of constants. Each of these constants may be preceded by an integer constant written m* which means the constant is specified m times. For example, to initially set all 400 array elements of the previously dimensioned array A(20,20) to zero, we could write

$$\text{DATA A}/400*0.0/$$

Unlabelled COMMON blocks cannot be initialized in this way. Note that the data statement can be used in any program segment in the manner discussed above, and is not confined to BLOCK DATA subprograms. It is good practice to use BLOCK DATA statements to initialize variables that will not be changed during execution. Examples of the use of these statements can be seen in later chapters.

Ordering of Specification Statements

The DIMENSION, COMMON, EQUIVALENCE, and DATA statements must precede any executable statements within a program or subprogram unit, and must, in Basic FORTRAN, appear in the order given, though in ASA FORTRAN the order is without significance.

Further FORTRAN Statements

TYPE statements

There are a number of non-executable FORTRAN **type** statements which are of the form

$$t \quad v_1, \ldots, v_n$$

where *t* is one of the following *type headings:* REAL, INTEGER, DOUBLE PRECISIO
COMPLEX or LOGICAL. Each v_i is a variable, array or function name, or an array
declaration. Type declarations are often used to overwrite the implicit type declaration
(integer variables starting with I,J,K,L,M or N, real variables with the other 20 alphabetic characters). Array declarations can also include DIMENSIONing information, in
which case a DIMENSION statement is not needed. Examples:

> REAL I1,J2,K3,B(200)
> INTEGER A1,A2,A3
> LOGICAL L1,L2,L3
> DOUBLE PRECISION BLOB(20)
> COMPLEX C(10,10)

EXTERNAL statements

If an argument in a subprogram call is a subprogram name or inbuilt *external* (that is,
out of line) function name (for example, SIN), this must be made known to the
compiler by a statement of the form

> EXTERNAL *name 1, name 2, . . . , name n*

where *name 1, . . . , name n* are the subprogram or function names. The EXTERNAL
statement appears in the calling program. For example, if we wanted to pass the name
of our FUNCTION subprogram ISTEP(defined on page (58) as an argument in a subprogram call to the FUNCTION subprogram IINT, and later we want to pass the inbuilt
external function SIN in a subprogram call to the SUBROUTINE subprogram GFUNC,
we would use the EXTERNAL statement as follows:

```
C    MAIN PROGRAM WITH FUNCTION SUBPROGRAMS ISTEP AND
C    IINT AND SUBROUTINE SUBPROGRAM GFUNC
     EXTERNAL ISTEP, SIN
     .
     .
     .
  25 IANS = IINT(ISTEP,1.0,J)
     .
     .
     .
  35 CALL GFUNC(1.5,SIN,K,L,M)
     .
     .
     .
```

The EXTERNAL statements tell the compiler that ISTEP, appearing in statement 25, and SIN, appearing in statement 35, is the name of a subprogram or external inbuilt function, rather than a variable or array name.

END statement

Every program unit (main program or subprogram) must terminate with a (possibly labelled) END statement.

STOP and PAUSE statements

There are only two FORTRAN statements whose effect constitutes interaction with the operator or operating system. They are the statements STOP and PAUSE, both of which may be followed by an optional octal constant. The STOP statement terminates the program, while the PAUSE statement causes the program to halt, awaiting operator or operating system intervention. For example, the loading of a tape. Operation may only be resumed by the operator or operating system.

Most manufacturers provide FORTRAN compilers which are a superset of ASA FORTRAN discussed here. The extensions range from the trivial, such as extending the number of alphameric characters in a variable name, to the profound, such as the provision of recursive subprogram calls. Nevertheless, it is recommended that FORTRAN programs be written in ASA FORTRAN in all but exceptional circumstances, since such programs should run with all compilers. If they don't, one then has the satisfaction of blaming the compiler writer rather than oneself!

Problems

3.1 Write a FORTRAN program segment to cyclically permute the K components of the array VEC. That is, VEC(2) is replaced by VEC(1), VEC(3) is replaced by VEC(2), and so on, with VEC(1) being replaced by VEC(K).

3.2 An array IMAT is 5 by 5, and IMAT(I,J) is I*J if I is even and J is odd, I+J if J is even and I is odd, and zero otherwise. Write FORTRAN statements to fill the array IMAT as specified.

3.3 A warehouse sells a number (not exceeding 200) of separate items. Each card in a deck of punched cards holds details of daily sales of an item for a month (20 days) and the selling price for that item. Maximum daily sales of any item do not exceed 800 units. An entry is included for each day and zero is punched when there are no sales on one day. Columns 1 to 60 are used for sales and columns 60 to 65 contain the price of the item in cents. The deck of cards is terminated with a card which has the value 999 punched in columns 1 to 3. The following incomplete program is intended to calculate and print out a sales summary, item by item.

```
C          COMMENTS BELONG HERE
           DIMENSION VALUE(200),NDEAD(200),AVERG(200),ISOLD(20)
C          SET UP LOGICAL UNIT NUMBERS FOR INSTALLATION
  1000  NPRNT = 3
           NREAD = 1
C          NOW INITIALIZE WHATEVER NEEDS TO BE INITIALIZED
  1100  ??
C          MAIN LOOP
  2000  READ(NREAD,2100)ISOLD,NCENT
  2100  FORMAT (??)
  2200  IF(ISOLD(1)–999)??
C          WE HAVE FOUND AN ITEM
  2300  ITEM = ?
C          CALCULATE NO. OF DEAD DAYS FOR THIS ITEM
  2400  NDEAD(ITEM) = KOUNT(ISOLD)
C          CALCULATE TOTAL NO. SOLD AND SALES VALUE FOR THE MONTH
  2600  ISUM = TOTAL(ISOLD)
  2700  VALUE(ITEM) = NCENT*ISUM/100.0
C          CALCULATE AND STORE AVERAGE DAILY SALES VALUE FOR
           THIS ITEM
  2800  ???
  3000  GO TO??
C
C          ALL ITEMS HAVE BEEN READ. NOW CALCULATE GRAND TOTAL
C          SALES AND PRINT OUT THE TABLE
C
  4000  NTOT = ITEM
  4100  GRAND = TOTME(VALUE,NTOT)
  4200  CALL PAGER (??)
C          LOOP FOR PRINTING OUT ITEM REPORT LINES
  5000  ITEM = ITEM+1
  5100  IF(ITEM–NTOT)5200,5200,6000
  5200  PRCNT = 100.00*VALUE(ITEM)/GRAND
C          NOW CHECK FOR BOTTOM OF PAGE
           LINES = LINES+2
  5300  IF(LINES–50)??
  5400  CALL PAGER (??)
  5600  WRITE (NPRNT,5700) ??
  5700  FORMAT (??)
           GO TO 5000
```

(cont.)

```
C        THAT IS ALL, FRIENDS
6000     STOP
         END
```

For each item, the table contains a line giving the item number within this summary and number of days on which no sales were made for that item, the average daily sales value (dollars and cents) for that item, and, for the month as a whole, the total sales values for this item as well as the percentage of the grand total sales value (for all items) represented by this item. The typical page of the table is to start as follows:

SALES REPORT		GRAND TOTAL SALES = $xxxxxxxxxx.xx	PAGE NO. xx	
ITEM	DEAD DAYS	AV. DAILY SALES	MONTH-VALUE	PERCENTAGE
xxx	xx	xxxxxxxx.xx	xxxxxxxx.xx	xx.xx
xxx	xx	xxxxxxxx.xx	xxxxxxxx.xx	xx.xx

. . .

Note that a page is assumed to have no more than 50 usable lines and blank lines appear between every two printed-out lines. The program must take care of this. The program on pages 68 and 69 has a number of pieces missing, indicated either by question marks or by the appearance of FUNCTION or SUBROUTINE subprograms which are not given explicitly. Study the program, then answer the following questions:

(a) Draw a summary flowchart for this program.

(b) From this flowchart, decide what needs to be initialized at statement number 1100 and write FORTRAN commands doing so.

(c) Supply suitable comment cards for the beginning of the program.

(d) Produce a flowchart and coding for FUNCTION KOUNT(ISOLD), see 2400.

(e) Produce a flowchart and coding for FUNCTION TOTAL(ISOLD), see 2600.

(f) Complete statements 2800 and 3000.

(g) Complete statements 5300 and 5600.

(h) By looking at the flowchart for the program as a whole, explain why a step which has been omitted by mistake has the result that no line at all will be printed out, other than a page heading for the first page. What is this missing step and where does it belong in the program?

(i) Produce a flowchart, but not full coding, for the SUBROUTINE PAGER, see statements 4200 and 5400. Make sure PAGER does all it has to do! What arguments must PAGER be given?

(j) Explain why the FUNCTION TOTME which appears in statement 4100 cannot be the same, nor can it have the same arguments, as FUNCTION TOTAL in statement 2600.

(This problem is from the Australian Computer Society draft entrance examination.)

3.4   Write a FORTRAN program to calculate

$$\text{erf}(1) = \frac{2}{\sqrt{\pi}} \int_0^1 e^{-t^2}\, dt$$

using Simpson's rule. The region of integration should be broken up into 100 steps.

3.5   Write a FORTRAN program to calculate the greatest common divisor of two integers using Euclid's algorithm.

3.6   Write a FORTRAN program to calculate and print out the first 100 prime numbers, given that the first two prime numbers are 2 and 3.

4. The ALGOL Language

4.1 Introduction

The first published version of ALGOL (ALGOrithmic Language) appeared in 1958 and was called, appropriately enough, ALGOL 58. It was designed by an international committee with the following terms of reference:

(1)   it should be as close as possible to standard mathematical notation and be readable with little further explanation;
(2)   it should be usable to describe computing processing in publications;
(3)   it should be mechanically translatable into machine language.

Upon publication of the language in 1958, a large amount of correspondence appeared in the literature pointing out shortcomings and difficulties in the language. In 1959 at the International Conference on Information Processing held in Paris, an important paper by J. Backus was presented. It contained a formal method of defining syntax – thereafter referred to as Backus Normal Form or BNF – and gave a definition of ALGOL in this form. This represented a more formal approach to language definition than that prevailing at the time. A brief description of BNF and an example of its use in defining a language is provided by the 'Revised report on the algorithmic language ALGOL 60', which can be found in virtually any book on ALGOL. The report contains a description of a considerably improved version of ALGOL 60, with most of the 'bugs' removed, and was issued in 1963. ALGOL 60 was itself a considerably improved version of ALGOL 58, and was issued as a result of an international committee meeting in Paris in 1960.

The language ALGOL 60 has received widespread acceptance by the European scientific community, though treated with more reserve by American scientists, many of whom tend to favour FORTRAN or PL/I. We will discuss and compare the two languages FORTRAN and ALGOL in the next chapter.

A further development of the language occurred in 1969 when the report of a committee headed by A. van Wijngaarden appeared. The language is known as ALGOL 68 and is not just a superset of ALGOL. It is a language with the same philosophy as ALGOL 60, but in its detailed definition it is quite different. At the time of writing, th only widely available compiler is the ALGOL 68-R system developed by the Royal Ra Establishment, Malvern, England, for the ICL 1900 series computers. The long-term future of this language is unclear, though it is most unlikely to replace ALGOL 60 in th short term. For these reasons we will only discuss ALGOL 60 in any detail, as defined in the 'Revised report on the algorithmic language ALGOL 60', which we will refer to simply as ALGOL.

4.2 Language Specification

Unlike FORTRAN, there are three distinct versions of the language. Firstly, there is th reference language, which is the defining language and working language of the commit tee. The 'Revised report' is written in this language. Secondly, there is the publication language, which is the language in which ALGOL programs may be written or publishe Finally, there is the hardware representation, which uses the character set of a particul computer installation, and is a condensed version of the reference language reflecting the limited number of characters available on standard input equipment.

An example of a function procedure to calculate the area of a circle given its radius is given below in two forms. Firstly, the publication language in which the programmer might write the program, and secondly, the hardware representation which is the form in which the procedure would be written onto coding sheets and would appear on punched cards.

(a) Publication language
real procedure area (r);
value r; **real** r;
area$:=\Pi_{\times}r^2$

(b) Hardware language
'REAL' 'PROCEDURE' AREA(R);
'VALUE' R;
'REAL' R;
AREA: = 3.141592654*R**2

Input and Output

One deficiency of ALGOL 60 is that input and output are not discussed at all in the ALGOL report. The language is designed to solve problems, with no consideration as to how numbers are to be input or output into/from the machine. Each manufacturer provides his own version of input and output procedures, which cause considerable problems when transferring a working program from one machine to another. An inter national standard for input/output was produced in 1964, but this has still not been officially approved and accepted. Some degree of standardization has been reached by common agreement between manufacturers. Later we will discuss two versions of input/output. Firstly, that of the ISO draft recommendation, as implemented on the

IBM 360 and CDC 6000 series computers, and secondly, the input/output procedures for ICL 1900 ALGOL.

The layout of a program on punched cards is extremely flexible. All 80 columns of a card may be used, with no specific role being adopted by any column. Cards may contain more than one statement, or one statement may continue for several cards. For this reason it is necessary to indicate the end of a statement (there are certain exceptions to this rule, which will be discussed later), and this is done by a semi-colon(;). The most common mistake made by programmers learning ALGOL is to leave out this symbol from one or more statements.

Character Set

The reference language contains a large character set, not all of which is available in the hardware language. The reference language character set is reproduced in the table below, along with the corresponding hardware representation (when it exists).

ALGOL publication language character	*ALGOL hardware representation character (card mode)*
$A, B, C, \ldots \ldots, X, Y, Z$	A,B,C, ... ,X,Y,Z
$a, b, c, \ldots \ldots, x, y, z$	Non-existent
$+, -, \times, /, \div, \uparrow$	$+, -, *, /, '/', \uparrow$ or **
$=, \neq, <$	= or 'EQ', # or 'NE', < or 'LT'
$\leqslant, >, \geqslant$	'LE', > or 'GT', 'GE'
$\equiv, \vee, \wedge$	'EQUIV', 'OR', 'AND'
$\neg, \supset, \sqcup$	'NOT', 'IMPL', % or '—'
( ) [ ] . ,	( ) [or '<'] or '>' . ,
$', ', :, ;, :=, _{10}$	'(', ')', :, ;, := or ←, & or '10'
go to, for, own, switch	'GO TO', 'FOR', 'OWN', 'SWITCH'
if, step, Boolean	'IF', 'STEP', 'BOOLEAN'
string, then, until	'STRING', 'THEN', 'UNTIL'
integer, label, else	'INTEGER', 'LABEL', 'ELSE'
while, real, value	'WHILE', 'REAL', 'VALUE'
begin, comment, array	'BEGIN', 'COMMENT', 'ARRAY'
procedure, end, do	'PROCEDURE', 'END', 'DO'

Though the last 22 characters (from **go to** to **do**) appear to be words, they are really considered single characters by the compiler. In the hardware representation it is possible to use REAL as a variable name, even though 'REAL' is a character. This is of course poor programming practice.

Constants

Integer constants are written as a string of digits with no other characters, such as commas, permitted. They may be preceded by a $+$ or $-$ sign.

Real constants are written as a string of digits containing exactly one internal decimal point. Alternatively, a real constant may be written as an integer or real consta followed by an exponent part, written $_{10}n$, with 'n' an integer, representing the power 10 by which the number is multiplied. An exponent part alone is also permitted. All real constants may be signed.

Logical constants: only two of these exist, written **true** or **false**.

Examples

Integer constants $1, -1342, 0, -769231$.

Real constants $+1.0, -134.2691, 0.0, -7.69231_{10}5$ $(= -769231.0)$, $125_{10}-4$ $(= 125.0 \times 10^{-4})$, $_{10}6$ $(= 1000000.0)$, $-001.53_{10}2$ $(= -153.0)$.

Note that an ALGOL real constant (unlike its FORTRAN counterpart) cannot termina with a decimal point, so that the following examples are illegal real constants: $15., 15._{10}5$. Note that there is no counterpart to FORTRAN's complex and double precision constants.

Variables

Variables are designated by identifier names which are strings of alphanumeric charact the first of which must be alphabetic (upper or lower case). No upper limit on the number of characters is specified, though a compiler limitation is often imposed. There is no predefined type convention, so that *all* identifier names must have their type de- clared by a **type declaration**. Subscripted variables are also permitted and any number of subscripts is allowed. Subscripts must be enclosed in *square* brackets, and can be complicated arithmetic expressions of **real** or **integer** type, or indeed, any expression which returns a numerical value. If a subscript is not an integer, it is implicitly rounde to the *nearest* integer. Internal blanks in an identifier name are ignored.

Examples

BLOB, FRED, blob, Fred, This is a valid name, var$[1,2,3.5,-6]$, array $[-\sin(x),\cos(x)+3.6_{10}1,-4]$, vec$[1,1.1,0,-9]$, two dimensional ARRAY$[9,-9]$, Fred$[$**if** $x>0$ **then** 4 **else** 8$]$.

Arithmetic Operators

There are six arithmetic operators, which are described in the table opposite.
Note that there are two distinct division operators. The real division operator $/$ gives a **real** result irrespective of the type of operands. The integer division operator $\div$ can on be used between operands of type **integer** and yields the largest integer whose absolut value is not greater than the absolute value of the true quotient. Thus $7 \div 4 = 1$, $-7 \div 4 =$ -1. The operators $+, -, \times,$ and $\uparrow$ yield **real** results unless the operands they connect are

Operator	Meaning	Examples
+	Addition	$X: = A+4.0$, $Y: = 4.0+3.0$
−	Subtraction	$X: = A−4.0$, $Y: = 4.0−3.0$
×	Multiplication	$X: = A×4.0$, $Y: = 4.0×3.0$
/	Real division	$X: = A/4.0$, $Y: = 4.0/3.0$
÷	Integer division	$X: = A÷4$, $Y: = 4÷3$
↑	Exponentiation	$X: = A↑4$, $Y: = 4.0↑3.0$

both of type **integer**. The exponent in an exponentiation operation must also be non-negative for a result of **integer** type to be obtained. In common with other languages, the raising of a negative, **real** constant or variable to a **real** exponent is undefined.

Relational Operators

The six standard relational operators are available in ALGOL and these are listed below:

Operator	Meaning	Example
>	greater than	**if** $A > B$ **then** $c:=$ test
⩾	greater than or equal to	**if** $A ⩾ B$ **then** $c:=$ test
=	equal to	**if** $A = B$ **then** $c:=$ test
≠	not equal to	**if** $A ≠ B$ **then** $c:=$ test
⩽	less than or equal to	**if** $A ⩽ B$ **then** $c:=$ test
<	less than	**if** $A < B$ **then** $c:=$ test

Relational operators are used to compare **real** or **integer** constants or variables. The usual dangers in comparing **real** variables for equality are, of course, present in ALGOL.

Boolean Operators

The five **Boolean** operators defined in the ALGOL report are described in the table below:

Operator	Meaning	Definition	Example
¬	not	¬B is **true/false** if B is **false/true**	**if** ¬B **then** go to A
∧	and	$B ∧ C$ is **true** if and only if B and C are both **true**. Otherwise it is **false**.	**if** $B ∧ C$ **then** go to A

Operator	Meaning	Definition	Examples
∨	or	$B \lor C$ is **true** unless B and C are both **false**, in which case it is **false**.	if $B \lor C$ then go to
⊃	implies	$B \supset C$ is **true** if B is **false** or if C is **true** or both. Otherwise it is **false**.	if $B \supset C$ then go to
≡	equivalent	$B \equiv C$ is **true** if B and C are both **true** or both **false**. Otherwise it is **false**.	if $B \equiv C$ then go to

In the above table, B and C are **Boolean** variables, while A is a statement label.

Hierarchy of Operators

The hierarchy of operators in ALGOL is similar to that in FORTRAN and is given below:

Operator or expression	Hierarchy	Order within hierarchy
Parenthesized expressions	10	inner to outer
↑	9	left to right
x / ÷	8	left to right
+ −	7	left to right
$< \leqslant = \geqslant > \neq$	6	left to right
¬ (not)	5	left to right
∧ (and)	4	left to right
∨ (or)	3	left to right
⊃ (implies)	2	left to right
≡ (equivalence)	1	left to right

Operators of highest hierarchical rank are evaluated first. Within a given hierarchy, operators are evaluated from left to right.

Standard Functions

The ALGOL report recommends the inclusion of the following standard functions. The are not mandatory, and some ALGOL compilers may offer a subset or a superset of the The full list is given opposite on page 77. (In the list, E is any arithmetic expression. The *type* of E may be **real** or **integer** but a **real** result will always be returned, except that *sign* and *entier* yield **integer** results.)

Standard function	Meaning	Example
abs(E)	absolute value of E	$y := \text{abs}(E)$
sign(E)	$\text{sign}(E) = \begin{cases} +1 \text{ if } E > 0 \\ 0 \text{ if } E = 0 \\ -1 \text{ if } E < 0 \end{cases}$	$y := \text{sign}(E) \times E$
sqrt(E)	positive square root of E	$y := \text{sqrt}(E)$
sin(E)	trigonometric sine of E	$y := \text{sin}(E)$
cos(E)	trigonometric cosine of E	$y := \text{cos}(E)$
arctan(E)	principal value of arctan(E) in radians in the range $$\left[-\frac{\pi}{2}, \frac{\pi}{2} \right]$$	$y := \text{arctan}(E)$
ln(E)	natural logarithm of E	$y := \text{ln}(E)$
exp(E)	exponential of E	$y := \text{exp}(E)$
entier(E)	gives the largest integer algebraically $\leq E$; e.g., entier($+1.6$) = 1 entier(-1.6) = -2	$iy := \text{entier}(E)$

These function names may be overwritten by *type declarations* so that the programmer may use *sin* as a valid identifier name — provided that it is not used as a function designator for *sin(x)* at the same time. It is clearly poor programming practice to do this except in exceptional circumstances.

Assignment Statements

These take the form v:=e where v is a variable name and e is an arithmetic expression. The symbol := is read 'becomes' or 'is replaced by' and is often written by the programmer as a left pointing arrow. This is simply because when writing by hand, ← flows more easily off the pen than does :=. This special symbol also helps overcome the conceptual difficulty faced by FORTRAN beginners when faced with FORTRAN statements like A = A + 1.0.

If v is of the type **real** and e is of the type **integer**, e is converted to **real** before assignment. Conversely, if v is of the type **integer** and e is of type **real**, then e is converted to the *nearest integer*[†] before assignment.

Multiple assignment is also permitted, so that expressions of the form

$$v_1 := v_2 := \ldots v_n := e$$

are valid and have the result that all the variables v_n are replaced by the value of e. In this case, all variables v_n must be of the same type.

[†]FORTRAN programmers, note well!

Multiple assignment of subscripted variables is permitted, but the order of assignment is often important. This takes place from left to right. For example, in the statement

$$\text{list } [k] := l := k := 5$$

k is first set to its initial value, say 9, then list $[9]$, l, and k are all assigned the value 5. Thus the statement above is equivalent to the three statements:

$$\text{list} [k] := 5;$$
$$l := 5;$$
$$k := 5.$$

Boolean assignment statements are also permitted and take the form

bv:=be

where **bv** is a **Boolean** variable and **be** is a **Boolean** expression. Some examples are:

(1) **B:=true**
(2) BOOLVAR:= $P \vee Q$
(3) IF:=$t < 0$

Here B Takes the value **true**, BOOLVAR is set to the truth value of $P \vee Q$, while IF is set to **true** if $t < 0$ and is **false** otherwise.

Statement Separators

Each statement must be separated from its successor by a semi-colon, though statements may posses any number of internal blanks, and may extend over several cards.

Examples

(1) $a:=b:=0;$ (2) $a:=$
  $a:=a+1;$ $b:=0; a:=a$
  $b:=b+a;$ $+1;$
                  $b \quad :=$
                  $b+ \quad a;$

(1) and (2) are fully equivalent, though the programmer who would write code like example (2) is undoubtedly certifiable!

Statement Labels

A statement may be labelled by an identifier, written before the statement to which it refers, and separated from it by a colon. Statements may also have more than one label

Examples

stat 1: $a:=a+1$;
A: $b:=b+1$;
Add 1: $c:=c+1$;
Label 1:   Label 2:   Label 3: $C:=4.0$;

The last statement may be referred to by any of its three labels.

The ALGOL report also allows statements to be labelled by unsigned integers — just like FORTRAN. This provision is not however available with a number of ALGOL compilers and for this reason its use is not recommended.

Compound Statements

An important feature of ALGOL and one that makes it so suitable for structured programming, is the provision for the formation of *compound statements* by writing a series of statements in *statement parentheses,* which are the characters **begin** and **end**. For example:

compl: **begin** $n:=0$;
           $n:=n+1$;
           $m:=n+k$
       end

is a compound statement and, like any other statement, may be labelled. The above compound statement has been labelled compl. Note that no semi-colon is needed before the **end** statement. Including a semi-colon before an **end** statement is not illegal, nor does it effect the results produced by the program, but it does usually lead to a less efficient object program, since it causes the compiler to generate a *dummy statement* — that is, a statement which does nothing.

Type Declarations

Since there is no predefined type convention in ALGOL, *all* variables must have their type declared before they are referenced. There are three type headings, given by the three ALGOL characters **real**, **integer**, and **Boolean**. Examples are:

real a,b,c,x1,x2;
integer i,j,j123,k;
Boolean $B1,B1$, Logical;

To declare subscripted variables it is necessary to include an **array** declaration in combination with a *type* declaration. In the absence of a type declaration, a **real array** is assumed. Some examples are:

Example	Meaning
real array $B[0:3]$	Storage must be allocated for a **real array** with 4 elements, $B[0],B[1],B[2]$, and $B[3]$.
real array $B[-0.2:3.4]$	Same meaning as above, array values are rounded to nearest integer.
integer array $i[1:5,1:8]$	Storage must be allocated for a 2 dimensional **integer array** with 40 elements $i[1,1],i[1,2],\ldots,i[5,8]$.
integer array $i[0.7:4.6,0.93:7.9]$	Same meaning as above.
integer array $mat[1:n,1:m-3*i,1:k[i]]$	This array is said to have *dynamic bounds*. The number of elements in this array is set during execution and depends on the value of the simple variables n and m, and the subscripted variable $k[i]$.
integer array intvec$[-7:9]$	One dimensional integer array of 17 elements, intvec$[-7]$,intvec$[-6]$, . . . , intvec$[9]$.

The position of type declarations within a program is of great significance and will be discussed later.

Blocks

A *block* is defined as a *compound statement* with at least one *type declaration*. All type declarations must appear between the **begin** character at the head of the block and the first executable statement of the block. The concept of a block and the role of block structure are of paramount importance in ALGOL and will be discussed more fully later

Program Control Statements

A variety of program control statements are available in ALGOL. These are used to vary the normal sequential execution of statements and are listed below:

go to statements
To branch unconditionally to a labelled statement a **go to** statement is used. It takes the form

$$\text{go to } sl$$

where sl is a statement label of an executable statement, or an expression that returns a value which is treated as a statement label, such as **go to** matrix $[2]$.

switch statements

A conditional **go to** may be formed by combining a **go to** statement with a **switch** declaration. A *switch declaration* is made at the beginning of the appropriate block and takes the form

$$\text{switch } s := sl_1, sl_2, sl_3, \ldots, sl_n$$

where s is a *switch identifier*, constrained by the usual rules for identifier names, and $sl_1, \ldots, sl_n$ is a *switch list*; that is, a list of statement labels, or expressions which take values which are treated as statement labels.

At some point in the block a **go to** statement of the form

$$\text{go to } s[e]$$

is written, where s is the switch identifier and e is an arithmetic expression. If the arithmetic expression is of type **real** it will be rounded to the *nearest* integer. If e takes the value n, control is transferred to the statement with label sl_n in the switch list. If sl_n is not in the switch list, no transfer takes place.

Example

$$\text{switch } switch1 := s1, s2, s3;$$
$$\vdots$$
$$\text{go to } switch1[i+2]$$

These statements transfer control to statement s1 if $i+2 = 1$, to statement s2 if $i+2 = 2$, and to statement s3 if $i+2 = 3$.

The most general form of a **go to** statement is obtained by combining a **go to** statement or *switch list* with a *conditional expression* (defined in the next section). This takes the form

$$\text{go to } conditional\ expression$$

or

$$\text{switch } sl := sl_1, sl_2, \ldots, sl_n$$

where the *switch list* $sl_1, \ldots, sl_n$ may contain not only statement labels, but also appropriate *conditional expressions*. This is best illustrated by some examples:

(1) **go to if** $M > N$ **then** $S1$ **else** $S2$;
(2) **go to if** $M > N$ **then** $S1$ **else if** $M = N$ **then** $S2$ **else** $S3$;
(3) **switch** $S := S1,$ **if** $X > 0$ **then** $S2$ **else** $S3$;

$$\vdots$$

$$\text{go to } S[i+k-m]$$

Example (1) is equivalent to **go to** $S1$ if $M > N$

                               go to $S2$ if $M \leqslant N$.

Example (2) is equivalent to **go to** $S1$ if $M > N$

                               go to $S2$ if $M = N$

                               go to $S3$ if $M < N$.

Example (3) is equivalent to **go to** $S1$ if $i+k-m = 1$

                               go to $S2$ if $i+k-m = 2$ and $X > 0$

                               go to $S3$ if $i+k-m = 2$ and $X \leqslant 0$.

Conditional Statements

We have already used a *conditional statement* without discussion in the previous section. In this section we will define and discuss the different forms of conditional statements.

if then statements
The simplest form of conditional statement takes the form

if be then S

where *be* is a *Boolean expression* and S is an *unconditional statement* or *simple arithmetic expression*. If the Boolean expression is **true**, statement S is executed, otherwise the statement following the **if** statement is executed. An unconditional statement is: (1) an assignment statement, or (2) a **go to** statement, or (3) a compound statement, or (4) a block, or (5) a dummy statement, or (6) a procedure statement. (Some of these statements have yet to be defined.) A simple arithmetic expression is an expression formed from numbers, variables, and function designators, possibly combined by arithmetic operators.

Some examples are:

if n < 0 then k:=−k;
if k < 0 then go to S1
if sin(X)+cos(X) ⩽ 0.5*exp(X) then go to less than;

Note that each **if** must be followed by a **then**, a fact that FORTRAN programmers tend to overlook.

An *unconditional statement* may also be a *compound statement,* so it is permissible to have such constructions as

if n < 0 then begin X:=X+1;
y:=sin(X);
z:=x+y
end;
S2: . . . . . .

if $n \geqslant 0$, control is immediately transferred to statement $S2$.

if then else statements

Another conditional statement provided in ALGOL is of the form

$$\text{if } be \text{ then } s_1 \text{ else } s_2$$

where be is a Boolean expression and s_1 is an unconditional statement or simple arithmetic expression. s_2 may, however, be any statement, conditional or unconditional, or any arithmetic expression. If the Boolean expression be is **true** then statement s_1 is executed, while if be is **false**, statement s_2 is executed. Note that the above statement could be replaced by the following sequence

> **if** be **then go to** $L1$;
> s_2;
> **go to** $L2$;
> $L1$: s_1;
> $L2$:   . . .

This sequence is so common however that ALGOL makes special provision for it in the much more compact and more readable, **if then else** statement.

Example

if $n < 0$ **then** $n{:=}n{+}1$ **else** $n{:=}n{\times}k$ the effect of which should be obvious.

Since s_2 may be *any* statement, it can be another conditional statement, including another **if then else** statement. Thus, such statements may be combined as in the following example:

> **if** $n < 0$ **then** set$:=0$ **else if** $m > 0$ **then** set$:=1$ **else** set$:=-1$

which defines set$:=0$ if $n < 0$
   set$:=1$ if $m > 0$
   set$:=-1$ otherwise.

At first sight there is an ambiguity here, as it is possible for $n < 0$ and $m > 0$, which corresponds to two different assignment statements for the variable *set*. There is a rule for evaluating such statements which removes the ambiguity, which may be stated as follows:

> Read from left to right until a **true** Boolean expression is encountered; the statement immediately following this expression is obeyed and the rest of the conditional statement is skipped. If all the Boolean expressions are **false**, the statement after the final **else** is obeyed.

In the previous example, if $n < 0$ and $m > 0$, the variable *set* then takes the value 0.

Conditional arithmetic statements are special cases of conditional statements of the form

$$\text{if } be \text{ then } A_1 \text{ else } A_2$$

where *be* is a Boolean expression, A_1 is a simple arithmetic expression and A_2 is any arithmetic expression. For example,

$$\text{if } A < B \text{ then } t:=1 \text{ else if } A = B \text{ then } t:= 0 \text{ else } t:=-1$$

These can be written in the following alternative form:

$$v:= \text{if } be \text{ then } c_1 \text{ else } c_2$$

where *v* is a valid identifier name, *be* is a Boolean expression, and c_1 and c_2 are valid right-hand sides of an assignment statement. For example, the above conditional arithmetic statement could be written

$$t:= \text{if } A < B \text{ then } +1 \text{ else if } A = B \text{ then } 0 \text{ else } -1;$$

In this form the character **else** is mandatory, so that the following statement is illegal:

$$t:= \text{if } A < B \text{ then } 1;$$

If a conditional arithmetic statement is parenthesized, it becomes a simple arithmetic statement, so that such statements as

$$z:= y+(\text{if } A < B \text{ then } 1 \text{ else } \text{sqrt}(A))$$

are valid. This statement sets $z \leftarrow y+1$ if $A < B$ and sets $z \leftarrow y+A^{\frac{1}{2}}$ if $A \geqslant B$.

Conditional Boolean statements are also defined by the ALGOL report, and take the form

$$\text{if } be_1 \text{ then } sbe \text{ else } be_2$$

where be_1 and be_2 are **Boolean** expressions and *sbe* is a simple **Boolean** expression. A simple Boolean expression differs from a Boolean expression in that the former may no be conditional.

Looping

Loops can be coded explicitly by using an **if** statement to control a loop counter; however, ALGOL has a special statement, a **for** statement, that allows loops to be coded implicitly.

The general form of the **for** statement is

$$\textbf{for } v:= flist \textbf{ do } s$$

where v is a *control variable*, *flist* is called a **for** *list*, and s is any executable statement. A **for** *list* may take a number of forms, which we discuss below in a sequence of increasing complexity.

(1) A **for** *list* may be a single number or a list of numbers representing values the control variable must take. For example

> **for** $i := 1$ **do** $y := y + x[i]$ sets $y = y + x[1]$
>
> **for** $i := 1,2,3,4,5$ **do** $y[i] := 3.14159265 \times i \times i$
>
> sets $y[1] = 3.14159265, y[2] = 3.14159265 \times 4, \ldots,$
>
> $y[5] = 3.14159265 \times 25.$

(2) Instead of numbers, arithmetic expressions may be used as **for** *list* elements. For example,

> **for** $i := x, x+y, x\uparrow 0.3 + 2.3xy$ **do begin** $m := m+1$;
>
> $y := r + i$;
>
> $s[m] := y \times i$
>
> **end**

In the above example the compound statement is executed three times: first with $i = x$, then with $i = x+y$, then with $i = x\uparrow 0.3 + 2.3xy$.

(3) The most common way to perform simple incrementing in a **for** *list* is to combine it with a **step until** element. This combination takes the form

> **for** $v :=$ *initial value* **step** *increment* **until** *final value* **do** S

For example, **for** $i := 1$ **step** 1 **until** 20 **do** $y[i] := i \times i$ stores the squares of the first twenty integers in array y.

The *initial value, increment*, and *final value* may be any arithmetic expression. Before execution of S, the value of the loop control variable is tested to see that it lies inside the defined range. For example, **for** $i := 1$ **step** 2 **until** 2 **do** S, would result in S being executed precisely once, with $i = 1$. On the other hand, **for** $i := 1$ **step** -2 **until** -3 **do** S would result in S being executed three times with $i = 1$, then with $i = -1$, then with $i = -3$, while **for** $i := 1$ **step** -2 **until** 2 **do** S would result in S not being executed at all.

(4) In order to perform a loop subject to some condition being satisfied, it is possible to use a **while** element. For example, we may wish to add the terms in a series until a term is reached which is less than 10^{-6}. This can be achieved with a **while** element as follows:

> $i := 0$; term$:= 1$; sum$:= 0$;
>
> **for** $i := i+1$ **while** term $\geqslant_{10} -6$ **do begin**
>
> term$:= \ldots$;
>
> sum$:=$sum$+$term
>
> **end**
>
> $\vdots$

In the above example, *i*, *sum*, and *term* must clearly be initialized before the **for** *list*. As an example, consider the Riemann zeta function $\zeta(3)$

$$\zeta(3) = \sum_{n=1}^{\infty} \frac{1}{n^3}$$

We will calculate this by adding successive terms until a term $<10^{-8}$ is found.

```
begin real term, zeta; integer n;
term:=1.0; zeta:=0.0;
for n:=1,1+n while term ≥ 10⁻⁸ do begin
                                term:=1/n↑3;
                                zeta:=zeta+term
                              end
end
```

In this example we have introduced a new concept, a *multi-element* **for** *list*. Such a **for** *list* comprises several of the **for** *lists* considered to date, separated by commas. Thus the **for** *list* above means that $n \leftarrow 1$ and the statement following the **do** is executed if term $\geq 10^{-8}$. *n* is then increased by 1 (so is set to 2) and the statement following the **do** executed again. *n* continues to be incremented by 1, followed by execution of the statement following the **do** as long as term $\geq 10^{-8}$.

A more complex example than that just given is the following:

for *j*:=0,5,10 **step** 3 **until** 15,17 **step** 4 **until** 38 **do** *S*

which causes *S* to be executed with *j* = 0,5,10,13,17,21,25,29,33,37 successively.

Nesting of **for** statements is permitted to any depth. Since the *range* of a **for** statement is the immediately following statement, it is not necessary to impose any rule about the range of nested loops (such as is necessary in FORTRAN). The structure of ALGOL takes this into account, as any sequences of nested **for** statements have appropriately restricted range.

It is permissible to use a **for** statement to follow the **then** of an **if then else** conditional statement. For example,

if $X > 0$ **then for** *y*:=1,2,3 **do** *z*[*y*] :=*y*+sqrt(*X*) **else** *z*[*y*] :=*t*×*X*

A counterpart to the FORTRAN CONTINUE statement is the ALGOL dummy statement. This is a labelled or unlabelled statement with text consisting of a string of blank characters. Thus

Label 1:

is a dummy statement which may be referenced by its label; e.g. **go to** Label 1. This dummy statement has, of course, no effect. It is useful as a point to transfer to. Placing a semicolon before an **end** character has the effect of generating an (unlabelled) dummy statement immediately before the **end** character.

Control Structures

By now the observant reader will have realized that most of our basic control structures have a direct counterpart in ALGOL. This is one of the reasons that ALGOL is a much better language for writing structured programs than FORTRAN or BASIC. Only the CASE structure has no direct counterpart in ALGOL, and is usually coded via a switch list, which is *not* a control structure leading to well-structured programs. However, the other basic control structures are readily representable in ALGOL. Concatenation is achieved by writing statements sequentially. The conditional clause IF C DO S is written in ALGOL as **if** C **then** S where S is any unconditional statement. In particular it can be a compound statement since the ALGOL language has statement parentheses, the **begin** and **end** characters. The alternative clause, written in natural language as IF C THEN S_1 ELSE S_2 is written in precisely the same way in ALGOL. The choice clause CASE(i) of $S_1,S_2,\ldots,S_N$ requires a switch identifier to be written at the head of the appropriate block as follows:

switch swident:= label1,label2, . . .,label$N$

then the CASE statement becomes

go to swident [i]

where label$j$, $\{j=1,N\}$ is the label of a section of code to handle case j. swident is the switch identifier. The repetition clause REPEAT S UNTIL C becomes

$$S;$$
if $7C$ **then** S

and the iteration clause WHILE C DO S becomes

iterate: **if** C **then begin** S;

go to iterate

end

or in a better structured form

for $V:=E$ **while** C **do** S

where V is a control variable, as described in the previous section. One or other of these equivalent structures will usually turn out to be more convenient. The second construction usually leads to a better structured program. The loop structure for $:=A$ TO B IN STEPS OF C DO S is written in ALGOL as **for** $i=A$ **step** C **until** B **do** S. Finally, the unconditional transfer control structure is written **go to** n, where n is a statement label. Bear in mind that this control structure need be used far less frequently in ALGOL than in FORTRAN or BASIC. Further, its frequent use in ALGOL is usually – though not inevitably – the hallmark of a badly structured program.

Block Structure

The concept of a block has been mentioned and its definition has already been given. Much greater consideration must be given to block structure however as it is one of the key structural points of the ALGOL language. Speaking extremely loosely, the *block structure* of an ALGOL program is the counterpart of the FORTRAN main program/ subprogram division. Intelligent use of the block structuring facility in ALGOL leads to highly modular programs. The manner in which variables are allocated within a block during program execution is also of considerable importance, as it contributes to the run time efficiency of the program. To understand these points better, let us develop the concept of a block from the simplest case.

A simple ALGOL program will comprise one block. A more complicated block structure would be an inner block contained within an outer block, or two inner blocks within an outer block. These structures are shown below:

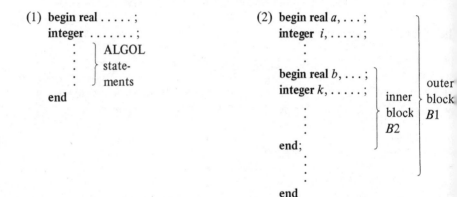

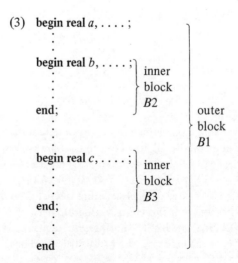

To appreciate the advantage of block structure it is necessary to understand the concept of the *scope* of a variable. An identifier appearing at the head of a block is available throughout the block, including any inner blocks. Outside the block, the variable is undefined. Further, and most importantly, any values assigned to variables or arrays declared in a block are *lost on exit from the block*. The *scope* of a variable is the set of statements over which an identifier is defined. A variable is said to be *local* to a block in which it is defined and *global* to any enclosed block which does not re-declare it.

All variables in example (1) are *local* to the whole program. In example (2) variables *a* and *i* are *local* to block *B1*, while being *global* to block *B2*. Variables *b* and *k* are *local* to block *B2* and undefined outside *B2*. In example (3) variable *a* is *local* to block *B1* and *global* to blocks *B2* and *B3*, while variable *b* is *local* to block *B2* and undefined outside block *B2*, and *c* is *local* to block *B3* and undefined outside block *B3*.

Since all variables are undefined on exit from a block, the compiler may reallocate the storage requirements of one block to another block. This usually results in considerable saving of storage space, as well as producing a more efficient object program. Thus at any particular time the program only uses as much storage as it needs at that time. The space occupied by an array in a block that is not the current block can be used to store some or all of the current block variable values. The local declaration of types within a block also leads to more readable programs.

If a variable is declared in both an outer and an inner block, it is said to be *re-declared* in the inner block. For example:

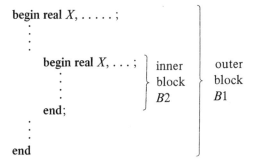

In this example *X* is re-declared in the inner block *B2*. *These are not the same variables* in the sense that the inner variable *X* has scope only over *B2*, while the outer variable *X* has scope only over those parts of *B1* which exclude *B2*. The value of *X* in *B1* is *suspended* on entry to block *B2*. On exit from *B2*, the value of *X* within *B2* is lost and the suspended value of *X* is re-assigned to it.

Block Entry and Exit

Blocks, like any other statement, may be labelled, as may statements within blocks. Labels are local to a block. A block may only be entered through its **begin** character –

that is, it is forbidden to transfer into the middle of a block from outside the block —
in much the same way that transferring into the range of a DO loop is forbidden in
FORTRAN. The reason for these restrictions should be obvious. Since variables are only
declared at the head of a block, these declaration statements must be encountered
first.

Example

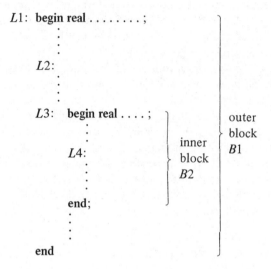

In this example, statements $L2$ and $L3$ are *local* to block $B1$ and *global* to block $B2$.
Label $L4$ is *local* to block $B2$ and undefined in that part of block $B1$ which excludes
$B2$.

Multiple Use of Identifiers

In ALGOL, the same identifier may be used for several different quantities — for
example, a **real** variable, an **integer** variable or a statement label — in different parts of
the program provided that the scopes of the quantities do not intersect. That is, the
quantities are required to have *disjoint scope*. This is best illustrated by two
examples:

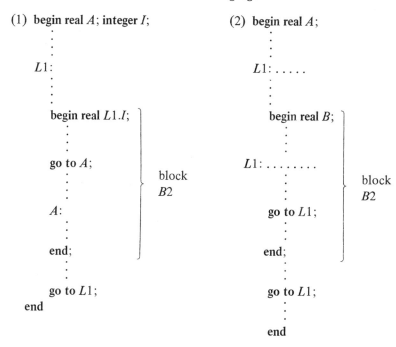

Both of these examples are legal ALGOL block structures. In the second example, the **go to** *L*1 in block *B*2 refers to statement *L*1 within *B*2, while the **go to** *L*1 outside block *B*2 refers to statement *L*1 outside block *B*2.

Own Variables

The fact that the value of variables is lost on exit from a block is sometimes undesirable. Thus ALGOL permits an **own** declaration, which takes the form

own *type list*

where *type* is one of **real, integer** or **Boolean** and *list* is a list of variables names. This facility allows the named variables to have the same value on re-entry to the block that they had on exit from the block. The use of variably dimensioned **own** arrays should be avoided, since the ALGOL report is insufficiently clear as to how such arrays will be handled.

Dynamic Bounds

If arrays with *dynamic bounds* are used, the block structure is essential, since the bound must be set in an enclosing block, so that the enclosed block has an array declaration with predefined bounds. For example:

```
begin integer lim;
      real ......;
           :
           :
           :
      lim:= ......;
           :
           :
      B2: begin real array mat[3:lim,4:lim];
           :                                          block
           :                                          B2
           end;
           :
           :
      end
```

On entry to block $B2$ it is necessary that the array bound *lim* has a pre-assigned valu
This is only possible with a nested block structure as shown since declarations must
appear at the head of a block.

Procedures

If a calculation is to be performed a number of times, it is usual to write a **procedur**
for the calculation. The procedure can then be invoked by the program whenever th
calculation is required. To define a procedure, a **procedure** *declaration* is used. A pr
cedure *declaration* comprises a **procedure** *heading,* giving the procedure name and a
information about the parameters of the procedure. This is followed by the coding
the procedure body, which must be a single statement — which, of course, includes
compound statement or a block. Like other declarations, **procedure** declarations m
appear at the head of a block. To use the procedure, a **procedure** *call* is made in the
of the program. The nature of the procedure call depends on the type of procedure
being called, just as in FORTRAN where FUNCTION subprograms and SUBROUT
subprograms are called differently.

A simple procedure declaration takes the form

procedure *name*; *S*

where *name* is the procedure name and *S* is the procedure body. For example a pro
cedure to calculate $\tan(X^2)$ might be written:

```
procedure tanXsq; begin
                  Xsq:=X×X;
                  t:=sin(Xsq)/cos(Xsq)
             end
```

This procedure declaration must be made within the scope of the variables X, Xsq and
. To call the procedure, its name must be given at the point in the program where it is
required. For example,

$$
\begin{array}{l}
\vdots \\
X:= 3.141592654/4.0; \\
\text{tan}X\text{sq}; \\
Z:= t/3.5 \\
\vdots
\end{array}
$$

As a result of this sequence of statements, t has the value returned by the procedure
with the particular value of X set before the procedure call; in this case, $\pi/4$.

In the previous example, since the variable Xsq was required only within the procedure, it could have been declared there; that is, the procedure could have been written

> **procedure** tan$X$sq; **begin real** Xsq;
> Xsq:=XxX;
> t:=sin(Xsq)/cos(Xsq)
> **end**

In which case the procedure statement is a *block*.

Procedures may also be written with formal parameters, so that the above example
could have been written in the following alternative form:

> **procedure** tan$X$sq(X, t);   **real** X,t;
> **begin real** Xsq; Xsq:=XxX;
> t:=sin(Xsq)/cos(Xsq)
> **end**

A procedure call in this case takes the form

$$
\begin{array}{l}
\vdots \\
\text{tan}X\text{sq}(3.141592654/4.0,q); \\
z:=q/3.5
\end{array}
$$

The parameters X and t in the procedure are dummy parameters, which are replaced by
the *name* of the parameters in the calling procedure. The type heading in the procedure
declaration is a statement to the compiler that the dummy parameters X and t will be
replaced by **real** parameters. It is *not* a *type* declaration for X and t. In the above example
X is referred to twice in the procedure definition. Each time it is referred to, the
quotient 3.141592654/4.0 must be calculated. We say that X *is called by name*. This is
the normal way procedures are called in ALGOL. If we require a procedure parameter
call to refer to the *value* of the parameter, this is referred to as *calling by value*. This has
the advantage that the value of the parameter is only calculated once, so that in the

above example the quotient would only have to be calculated once. It is usually more efficient to call procedure parameters by value. To do this, a **value** declaration must be included in the assigned procedure call. Rewriting this example yet again, this time including a **value** declaration, we have

> **procedure** $\tan X\mathrm{sq}(X,t)$; **value** X; **real** X,t;
> **begin real** $X\mathrm{sq}$; $X\mathrm{sq}:=X{\times}X$;
> $t:=\sin(X\mathrm{sq})/\cos(X\mathrm{sq})$
> **end**

The parameters in a procedure call must correspond in number to those listed in the procedure heading. The formal parameters following the procedure name may be variables, arrays, labels or other procedure identifiers. The identifiers of formal parameters are undefined outside the procedure block.

A general procedure to sum the first $n+1$ terms of a sequence $\sum_{i=0}^{n} t(i)$, might be written

> **procedure** $\mathrm{sum}(ans,n,term,i)$; **real** $ans,term$;
>   **integer** n,i; **begin** $ans:=0$;
>   **for** $i=0$ **step** 1 **until** n **do**
> $ans:=ans+term$
>   **end**

and the procedure may be called by

(1) $\mathrm{sum}(f1,100,1/(i+1),i)$; then $f1= \sum_{i=0}^{100} 1/(i+1)$

(2) $\mathrm{sum}(f2,30,i{\times}\sin(i),i)$; then $f2 = \sum_{i=0}^{30} i{\times}\sin(i)$

(3) $\mathrm{sum}(f3,20,a[i]+a[2{\times}i],i)$; then $f3 = \sum_{i=0}^{20} \{a[i]+a[2i]\}$

where $a[i]$ is a one dimensional array.

This example illustrates a programming technique sometimes known as 'Jensen's device'. This consists of a **for** statement within a procedure, with the controlled variable a formal parameter called by name.

Function Procedures

A function procedure is usually used when only a single result is to be returned. It is written as a **procedure**, with the **procedure** heading preceded by a type declaration

integer, real or **Boolean**, which is the *type* of the result produced. The *procedure name* is used for the variable to which the procedure result is assigned. To calculate the tangent of X^2 using a function procedure, one might write the following:

> **real procedure** tansq(X); **value** X; **real** X;
>   **begin real** Xsq; Xsq:=XxX;
>   tansq:=sin(Xsq)/cos(Xsq)
>   **end**

The procedure is called by naming the procedure on the right-hand side of an assignment statement. Example:

> :
> :
> z:= tansq(3.141592654/4.0)/3.5
> :
> :

Recursive Procedures

Unlike FORTRAN, ALGOL allows a procedure to call itself. This is called *recursion* and is a highly vaunted feature of ALGOL. The most common example of a recursive program is a program to calculate $n!$ from the definition $n! = n(n-1)!$ and $1! = 1$. An **integer** procedure to calculate $n!$ in this way is the following:

> **integer procedure** $n$fac(n); **value** n; **integer** n;
>   **if** $n = 1$ **then** $n$fac:=1 **else**
> $n$fac:=$n$fac($n-1$)xn;

Note that this is an incredibly inefficient way to calculate $n!$, but it suffices to illustrate a *recursive procedure*. The efficiency of recursive procedures will be discussed more fully in the next chapter.

Procedure Parameter Delimiters

In order to remind the programmer of the meaning of arguments in a procedure parameter list, ALGOL permits a second parameter delimiter in addition to the more usual comma. It consists of a right parenthesis, followed by a string of letters, followed by a colon and a left parenthesis.

Examples
(1)) this is a parameter delimiter :(
(2)) this is fully equivalent to a comma :(

This might be used in the procedure heading for procedure sum on page 94, where the heading **procedure** sum(ans,n,term,i) could be replaced by

procedure sum(ans) is the sum of: (n) terms of : (term) with argument : (i)

which makes the purpose of the procedure much more obvious to anyone reading the program. This facility does not, of course, add any power to the ALGOL language, but makes for ease of program interpretation.

Comments and Strings

There are a variety of ways of inserting comments into an ALGOL program. These comments do not affect the meaning of the program, but make it clear to the reader just what is going on. This at least *should* be the purpose of comments! One way of inserting a comment is the alternative form of parameter delimiter discussed in the previous section. Another way is to use the ALGOL **comment** character. A **comment** following a semi-colon or a **begin** may be followed by a sequence of characters terminated by a semi-colon. This sequence of characters is comment.

Example

> **comment** calculate required sums here;
> **begin comment** this is a comment;

Finally, any sequence of characters following an **end** and not containing a semi-colon **end** or **else** is a comment. Thus

> . . . **end** this is a comment **else** . . .
> . . . **end** this too is a comment; . . .
> . . . **end** even this is a comment **end** . . .

This is a common trap for young ALGOL players. A semi-colon is inadvertently omitted after an **end** character, and the next statement is interpreted not as an executable statement, but as a comment.

An ALGOL string is useful when one wishes to print out headings or other alphameric information. An ALGOL string is a string of characters contained within string quotes, which are ' and '. Spaces in a string must be indicated by the space the characters character.

Examples

> 'THIS⌴IS⌴A⌴STRING'
> 'MONDAY⌴APRIL⌴5TH'
> 'RING⌴A⌴DING⌴I⌴AM⌴A⌴STRING'

Other Algol Characters

The two ALGOL characters **string** and **label** have yet to be discussed. These characters are both declarators, and therefore must appear at the head of a block. A list of variable names follows these declarators, and these variables are defined as labels or strings. No

alculation can be performed on **label** or **string** variables. The **label** declarator enables a
tatement label to be used as an actual parameter. An example of the use of the **string**
eclarator is given in the next section on input/output. Note that strings and labels do
ot have to be declared in general, unlike real, integer or Boolean variables, but only if
hey are going to be used as parameters in a procedure call.

This covers most of the material contain in the revised ALGOL 60 report. For the
ormal definition of ALGOL language features, the report should be referred to,
hough the material in this chapter should provide sufficient information in most
nstances.

nput and Output

Finally, we will discuss the question of input and output which is not discussed in the
ALGOL report. Input/output facilities vary from manufacturer to manufacturer, though
a number of manufacturers provide the procedures specified in the ISO draft standard,
with variations ranging from minor to major. ALGOL, as implemented on the CDC 6000
eries and IBM 360 series, is close to the ISO draft standard, while ICL input/output
ears no resemblance to the draft standard whatsoever.

The ISO draft standard recommends seven basic input/output procedures; namely;

inreal	(c,rv)
outreal	(c,re)
inarray	(c,rai)
outarray	(c,rai)
insymbol	$(c,string,iv)$
outsymbol	$(c,string,iv)$
length	$(string)$

where c is a channel number and refers to a peripheral device, rv is a real variable, re is a
eal expression, rai is a real array identifier, *string* is a string and must be in string
uotes, while iv is an integer variable.

Examples

1) inreal $(3,A)$ — reads a number from peripheral device 3 and assigns its value to A.

2) outreal $(6,A\uparrow2+\sin(B))$ — evaluates $A^2+\sin(B)$ and outputs its value on device 6.

3) inarray $(4,A)$
   outarray $(4,A)$ — reads/writes the elements of array A in dictionary order from/onto device 4.

4) outsymbol $(2, `ABC',K)$ — writes one letter on device 2. The letter is A if $K = 1$, B if $K = 2$, and C if $K = 3$.

5) insymbol $(3, `ABC',K)$ — reads a symbol from device 3. If the symbol is not one of A,B or C, then K is given the value 0. If it is A,B or C, then K is given the value 1,2 or 3 respectively.

6) length $(`ABC')$ — counts the number of symbols in a string; in this case 3.

To see how these procedures may be used to print out five real variables, preceded by a heading, consider the following example:

```
begin real data, x,y,z,a,b;
        procedure list (string); string string;
        begin integer n;
        for n:=1 step 1 until length (string) do outsymbol (3,string,n)
        end;
          :
          :
          :
        list ('x ⌊⌋ ⌊⌋y⌊⌋ ⌊⌋z ⌊⌋ ⌊⌋a⌊⌋ ⌊⌋b');
          :
          :
          :
        for data:=x,y,z,a,b do outreal (3,data)
          :
          :
    end
```

This will output the heading *x y z a b* followed by the numerical values of the real variables x,y,z,a, and b. The **string** declaration has been introduced here and its purpose is to state that the identifier(s) following the **string** declaration are strings. No calculations may be performed on **string** variables.

ICL 1900 Series Input and Output

Input: To read numbers from a card, one may use a *read* instruction. This takes the form

$$v:=read;$$

where v is a variable name. The next number on a card is then read and assigned to variable v. There is no restriction on the format of numbers on cards, and the following delimiters denote the end of a number:

(1) double space,
(2) comma,
(3) semi-colon,
(4) new card.

Numbers are read according to the *type* of the variable v in the read statement. If v has been declared an *integer* then the number read is converted to integer mode before assignment. *read* statements can also be included in arithmetic expressions, as in

(1) $x:=y+$read
(2) if $x > 0$ then $y:=k$ else $y:=$read$+k$

In the above examples, *read* is replaced by the next number on the input file before assignment.

Output: The usual output instruction is

$$\text{print}(v,n,m)$$

where *v* is any variable or arithmetic expression, and *n* and *m* are integer constants. The integer constants control the format of the output. The above instruction causes the present value of *v* to be printed as a fixed point, floating point or integral number, depending on the values of *n* and *m*, according to the following rules:

1) $n \leqslant 12, m \leqslant 12$
2) if $m = 0$ an *n* digit integer is printed out, possibly preceded by a minus sign and followed by two blanks. Thus $n+3$ characters are used by this field.
3) if $n = 0$ an $m+1$ digit floating point number is printed, of the form
   $\pm x.xxxx \ldots xx\&\pm xx$ followed by two spaces. Thus, this field uses $m+9$ characters.
   $\underbrace{\qquad\qquad}$
   m digits
4) if $m \neq 0$ and $n \neq 0$ a fixed point number with *n* digits before the decimal point and *m* digits after the decimal point is printed, followed by four spaces. Thus $m+n+6$ characters are used by this field.

The following instructions are available to control the printer:

1) newline (*i*); where *i* is an integer constant. This instruction causes a skip of *i* lines.
2) paperthrow; produces a skip to the head of a new page.
3) space (*n*); where *n* is an integer constant. This instruction causes a horizontal displacement of *n* spaces.

To print out headings, a *write text* instruction is used, which takes the form:

$$\text{write text } (string)$$

where *string* is any string, as previously defined. The three layout characters *s,c,* and *p* may be included within a string, and should themselves be in string quotes, as in write text ("*p*'ANSWERS'2*c*");. The effect of layout characters is shown in the table below.

Layout character	Meaning
ns	skip *n* spaces. If *n* is absent, skip one space.
nc	skip *n* lines. If *n* is absent, skip one line.
np	skip *n* pages. If *n* is absent, skip one page.

The foregoing are only some of the input/output facilities of ICL 1900 ALGOL. For fuller details, the manual appropriate to your installation should be consulted.

Problems

4.1 Write ALGOL arithmetic expressions corresponding to the following expressions given in ordinary mathematical notation. All quantities can be assumed to be of type **real**:

(a) $\dfrac{A+4}{2B}$,   (b) A^{j+1},   (c) $\dfrac{(A+B)(C+D)}{E+F}$,   (d) $\log(y+\sqrt{(y^2-1)})$.

4.2 Write ALGOL assignment statements to perform the following operations, assuming all quantities to be of type **real**:

(a) $p = \left(\dfrac{A}{A+1.5}\right)^{3.5}$,   (b) $GC = \dfrac{h}{4} \cdot \dfrac{4R-h}{3R-h}$,   (c) Set l and $m = \dfrac{a^2 + \dfrac{1}{(c-d)^2}}{b^3+(c+a)^2 e}$

4.3 What will be the final values of the variables in the following program segment after the last statement has been executed?

```
integer      a,b;
             a:= −6;
             b:= −2;
             CC: if a < b then go to AA else go to BB;
        AA:a:= a+5;
             go to CC;
        BB: if a < 0 then a:= a+3;
             b:= a+b;
             if b ≤ 0 then go to CC;
        end: b:= a−b
```

4.4 Write ALGOL statements:
(a) to place the algebraically largest of X, Y, and Z in MAX,
(b) to compute the value of FUN defined by

$$FUN = \begin{cases} -2 & x < -2 \\ \dfrac{x}{2}-1 & -2 \leqslant x \leqslant -1 \\ \dfrac{5x}{2}+x^2 & -1 \leqslant x \leqslant 0 \\ 0 & x > 0 \end{cases}$$

4.5 Write a block to evaluate

$$\sum_{r=1}^{\infty} \frac{1}{(2^r+1)^{4/3}}$$

neglecting all terms less than 10^{-8}.

4.6 Write statements to form the arrays defined by

(a) $T_n = \dfrac{2n+1}{(2n)!}$ $\quad n = 1, 2, \ldots, p$

(b) $T_{q,s} = (1+r^2)(rs-2)$ $\quad r = -10, -8, \ldots, 10$
$$s = 0, 1, \ldots, 5$$
$$q = r/2$$

4.7 Write a procedure declaration for a procedure which, given the lengths a, b, c of the three sides of a plane triangle, evaluates the area of the triangle.

4.8 Write the programs specified at the end of the chapter on BASIC, page 42, in ALGOL.

4.9 The figure below shows a *latin square* of order 4. Each row and column contains the digits 1 to 4 with no repetition:

2	4	3	1
3	2	1	4
1	3	4	2
4	1	2	3

The following construction can be used to calculate latin squares of any order n. We will illustrate this for the case $n = 5$.

(1) Obtain the first row (or column) as any permutation of the numbers 1 to 5, for example 3 5 4 1 2.

(2) The first row enables us to calculate all the other rows as follows. The first element in the first row is 3. Element number 3 in the first row is 4. Write down the number 4 in the 5 successive rows in columns 3,5,4,1 and 2 respectively, as shown below.

3	5	4	1	2
				4
		4		
4				
	4			

The next cyclic permutation of the first row is 5 4 1 2 3, which refers to the placing of the 5th element; i.e. 2, into the 5 rows. This process is continued until the latin square is filled, which then looks like this:

$$3 \quad 5 \quad 4 \quad 1 \quad 2$$
$$1 \quad 3 \quad 5 \quad 2 \quad 4$$
$$2 \quad 1 \quad 3 \quad 4 \quad 5$$
$$4 \quad 2 \quad 1 \quad 5 \quad 3$$
$$5 \quad 4 \quad 2 \quad 3 \quad 1$$

Write an **ALGOL** program to read from a card

(a)   the order n of the latin square and
(b)   the first row, and then to calculate and print out the complete latin square.

A fresh set of data should be read, and the process continued until a zero is read as the order of the square. Note that there are other — simpler — ways of constructing a latin square, but doing it this way is a good exercise.

5. FORTRAN *vs* ALGOL

5.1 Comparisons

In discussing the relative merits of the two most commonly used high-level scientific languages, FORTRAN and ALGOL, it should be made clear at the outset that it is not possible to conclude that one is unequivocally superior to the other. Both of them have certain advantages in certain situations. If one has a clear understanding of the relative merits and shortcomings of the two languages, it is then easier to decide in a given situation which language is more suitable.

In comparing the two languages, the following points should be borne in mind.

(1) By the mere existence of a large number of users of both languages, the investment in software and expertise is such that neither language is likely to be discarded in favour of the other in the foreseeable future.

(2) ALGOL is undoubtedly a more modern and better planned language than FORTRAN, and is a far superior language in which to write well-structured, modular programs. This is a consequence of the wider variety of control structures directly available in ALGOL, and the block-structuring facility of ALGOL.

(3) Stylistically, the presence of statement parentheses and hence compound statements in ALGOL can lead to more readable programs.

(4) FORTRAN is defined in normal prose, whereas the ALGOL report is written in the more rigorous *Backus normal form.* Nevertheless, because ALGOL is a more complex language, there is more incompatibility due to misunderstanding of the language definition in ALGOL than in FORTRAN.

(5) Both languages are good for formulating and solving a wide range of numerical and scientific problems.

(6) Both languages make very heavy weather of business data processing, list processing, and string manipulation. This is possibly an unfair criticism as neither language was designed with these aims in view.

(7) ALGOL programs are normally less portable than FORTRAN programs due to the different hardware representation and character sets in different environments.

(8) Input/output facilities of FORTRAN are better than those of ALGOL. This is accentuated by the lack of standardization of input/output facilities in ALGOL.

(9) There are a number of variations of FORTRAN that sometimes prevent a program written for one compiler from running under another compiler. This difficulty can be minimized by sticking to ASA standard FORTRAN, which should run on most FORTRAN compilers.

(10) Many ALGOL compilers do not contain *all* the features of the language specified in the ALGOL report. This introduces a certain amount of machine dependence into ALGOL programs.

(11) There are more exclusions and restrictions in the FORTRAN language than in ALGOL.

(12) ALGOL, being a more complex language, takes longer to learn than FORTRAN. On the other hand, it is possible to learn a FORTRAN-like subset of ALGOL in about the same length of time that it takes to learn FORTRAN.

(13) ALGOL is a problem statement, problem solving language, while FORTRAN is a machine oriented language.

(14) Both languages have been designed for a batch processing environment.

(15) ALGOL has three language forms: the reference language, the hardware language, and the publication language. FORTRAN has only one language form. As a result, ALGOL is a better publication language for algorithms.

(16) The absence of a subroutine facility in ALGOL means that it is more difficult to incorporate a section of code written by another programmer into a program. That is, the interface between ALGOL blocks needs more careful planning than does the interface between FORTRAN subroutines.

(17) The block structure of ALGOL provides for dynamic allocation of storage, resulting in more economical use of storage during execution. Equivalent economy in FORTRAN requires careful programming, making use of COMMON storage and EQUIVALENCED variables wherever possible.

(18) In FORTRAN, typing or punching an erroneous variable name may not be detected by the compiler, whereas in ALGOL it probably would be. For example, BLOB = BL0B+1.0 in FORTRAN, where in the second reference to BLOB a zero has been accidentally punched instead of the letter oh, could cause a difficult to trace error. In ALGOL, it is likely that only the correct form of variable name would be listed in the type declaration statement at the head of the block, so that given the statement BLOB = BL0B+1.0, the compiler would detect an error and print a message to the effect that the variable BLOB was of unspecified type.

(19) No complex or double-precision constants exist in ALGOL 60, which is often a nuisance in scientific calculations.

(20) In ALGOL, procedure parameters can be passed in two ways, by name and by value. In FORTRAN all parameters are passed by location. That is, the address of the parameter is passed.

(21) Partly as a result of being a more complex language than FORTRAN, ALGOL

fails to make use of its full potential. In particular, most ALGOL compilers will accept and handle correctly a number of natural extensions of the language.

(22) The superior control structures of ALGOL allow an ALGOL program to be more readily reduced to a form in which a proof of program correctness can be easily given.

(23) The provision of *dynamic bounds* in ALGOL is a very important feature that FORTRAN lacks. It permits more efficient use of storage in those frequent situations where the size of an array is not known in advance.

(24) The presence of recursive procedures in ALGOL means that it is more suitable for certain computing processes that are naturally formulated recursively. FORTRAN can be used for this purpose, but only with some difficulty. However recursion is an important concept, and the absence of a recursive subroutine facility in FORTRAN is often keenly felt. In the next section we shall show how such functions can be coded in FORTRAN using only the existing language facilities.

Many of the criticisms of ALGOL 60 implied by the above comments have been answered by ALGOL 68 and by ALGOL–W. In ALGOL 68 this has been achieved at a cost of greatly-increased language complexity, while in ALGOL–W a number of restrictions and extensions have answered the bulk of the criticisms.

5.2   Recursive Functions

The standard example of a recursively defined function is the factorial function $n!$. This can be defined *iteratively* as

$$n! = n(n-1)(n-2)\ldots 3.2.1.$$

or *recursively* as

$$n! = n(n-1)! \text{ and } 0! = 1.$$

In the iterative definition it is usual to use a string of dots to represent some obvious intermediate steps, while in the recursive definition the function is expressed in terms of itself, with an auxiliary condition to prevent the recursive definition from going on forever.

To evaluate the factorial function from its iterative definition is quite straightforward. One merely sets up a loop with the loop variable ranging from 2 to N, initializes the function with the value 1, and multiplies repeatedly by the loop variable. To evaluate the function from its recursive definition is also quite straightforward in ALGOL, as the following procedure shows:

```
integer procedure fact(n); value n; integer n;
fact:=if n=0 then 1 else n×fact(n−1);
```

However, an equivalent sequence of FORTRAN statements, as in the following subprogram, is quite illegal:

```
                    INTEGER FUNCTION FACT(N)
                    IF (N.GT.0) GO TO 10
                    FACT = 1
                    RETURN
                10  FACT = FACT(N-1)*N
                    RETURN
                    END
```

To explain why this is illegal in FORTRAN, it is necessary to understand how subprograms are linked. When a subprogram is called, its *return address* is stored in a regist
That is, the RETURN statement effects a transfer to the return address when it is encountered. A second subroutine call to the same subroutine, before the RETURN state
ment is encountered, would wipe out the original return address and replace it with the
new return address. Thus the subprogram could never return to its original return
address. This difficulty is overcome in ALGOL by the provision of a *stack* to hold all tl
return addresses. A *stack*, sometimes called a *push-down list* or a *first-in last-out queue,*
is just a pile of objects, in this case return addresses, which are accessed on a first-in
last-out basis.

Before considering how to overcome this problem in FORTRAN, we should say a
little about the usefulness of recursion. Some ALGOL champions consider that the
provision of recursion gives ALGOL so great an advantage over FORTRAN as to be
positively overwhelming. On the other hand, some champions of FORTRAN argue tha
recursion is very rarely used, and when it is used, the problem could probably be solvec
iteratively in a more efficient manner (this is certainly true for the factorial function
just considered). As is so often the case when widely disparate views are held, the truth
probably lies somewhere in between. Recursion is useful in many applications, notably
list processing, syntactic analysis, compiler writing, formula manipulation, and certain
sorting applications. However, the indiscriminate use of recursive definitions in programming accounts for a lot of run time inefficiencies, since there are hidden costs
involved in using a recursive facility. These costs include the storage needed by the stac
and the additional compilation time involved in handling recursively-defined functions.
Nevertheless, recursion is sufficiently important that FORTRAN programmers should
be able to program recursive functions.

The key idea in doing this is quite simple. The programmer provides a *stack,* which
doesn't store the return addresses, since these are not known, but rather the arguments
for each function call. The value of the function can then be constructed from the stac
once the terminating conditions are satisfied. This scheme is illustrated in outline in
Figure 5.1.

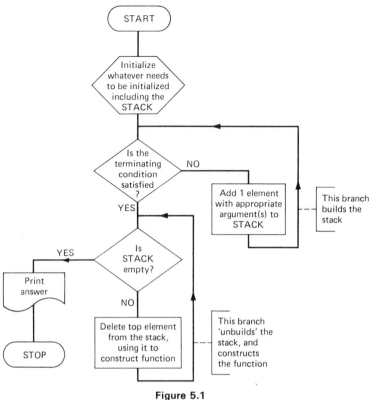

Figure 5.1

After initialization there are two branches: one to build the stack, with appropriate arguments to construct the function, and another to construct the function from the stack, once the stack is full.

We will now illustrate this technique with a number of examples. Consider first the factorial function, N!. The stack is a one-dimensional array S of dimension N. S(K) stores the Kth element in the stack, and when the stack is full, the N elements store the values N, (N−1), (N−2), (N−3), . . . , 1. These are then multiplied together in the 'unstack' branch. The flowchart (Figure 5.2) and corresponding FORTRAN program are shown on page 108.

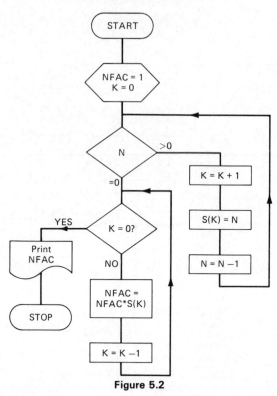

Figure 5.2

The FORTRAN program is written as a FUNCTION subprogram of type real, since N! stored as an integer overflows the word length on most computers for quite small values of N.

```
      FUNCTION XFACK(S,N)
      DIMENSION S(N)
      XFACK= 1.0
      K = 0
      XN = N
   5  IF (XN .LT. 0.001) GO TO 10
      K = K+1
      S(K) = XN
      XN = XN−1.0
      GO TO 5
  10  IF (K .GT. 0) GO TO 40
      RETURN
  40  XFACK = XFACK*S(K)
      K = K−1
      GO TO 10
      END
```

5.3 Ackerman's Function

A less trivial example is provided by Ackerman's function. The principal purpose of thi unction is to illustrate the nature of recursive function definition. It is defined by

$$A(0,n) = n+1$$
$$A(m,0) = A(m-1,1)$$
$$A(m,n) = A(m-1,A(m,n-1))$$

with m and n non-negative integers. An ALGOL program to calculate this function is hown below:

```
begin integer m,n,ans;
      integer procedure acker(m,n); value m,n; integer m,n;
         begin
            if m = 0 then acker:=n+1 else
            if n = 0 then acker:=acker(m-1,1) else
               acker:=acker(m-1, acker(m,n-1))
         end   now set the required arguments of
               ackerman's function;
      m:= ... ;
      n:= ... ;
      ans:=acker(m,n); comment now print this out using the
            output routine provided at your installation;
      print(ans,12,0)
end
```

The method of inputing the arguments m and n has been left undefined. A specific value can be set by an assignment statement, or the arguments can be read in from an input device. Making use of a recursive procedure call, the program is simplicity itself to write.

Before writing a FORTRAN program to calculate this function, it is instructive to consider an example of its evaluation. Taking $A(2,2)$ as a simple non-trivial case, we find from the definition:

$$
\begin{aligned}
A(2,2) &= A(1,A(2,1)) = A(1,A(1,A(2,0))) = A(1,A(1,A(1,1))) \\
&= A(1,A(1,A(0,A(1,0)))) = A(1,A(1,A(0,A(0,1)))) = A(1,A(1,A(0,2))) \\
&= A(1,A(1,3)) = A(1,A(0,A(1,2))) = A(1,A(0,A(0,A(1,1)))) \\
&= A(1,A(0,A(0,A(0,A(1,0))))) = A(1,A(0,A(0,A(0,A(0,1))))) \\
&= A(1,A(0,A(0,A(0,2)))) = A(1,A(0,A(0,3))) = A(1,A(0,4)) = A(1,5) \\
&= A(0,A(1,4)) = A(0,A(0,A(1,3))) = A(0,A(0,A(0,A(1,2)))) \\
&= A(0,A(0,A(0,A(0,A(1,1))))) = A(0,A(0,A(0,A(0,A(0,A(1,0)))))) \\
&= A(0,A(0,A(0,A(0,A(0,A(0,1)))))) = A(0,A(0,A(0,A(0,A(0,2))))) \\
&= A(0,A(0,A(0,A(0,3)))) = A(0,A(0,A(0,4))) = A(0,A(0,5)) \\
&= A(0,6) = 7.
\end{aligned}
$$

Rather than take fright at this rather formidable array of symbols, let us examine the structure of this expression. At each step the function is nested to a certain depth k; that is,

$$A(n_0,A(n_1,A(n_2,A(n_3, \ldots A(n_{k-1},A(n_k,p))) \ldots)$$

At the next step, the function is nested to a depth of $k+1$ or $k-1$. For our stack we nee
an indicator IND = k to record the depth to which the As are nested, while in the stack
we record IND elements, which are the values represented by $n_0, n_1, \ldots, n_{k-1}$. In the
current value of M we store n_k, and in the current value of N we store p. In going from
one step to the next, the stack size (and thus IND) is increased or decreased by one.
Therefore M and N are changed, while n_{k-1} is deleted from the stack (if the stack is
being decreased by one) or a new element is added to the stack (if the stack is being
increased by one). The flowchart is then as shown in Figure 5.3.

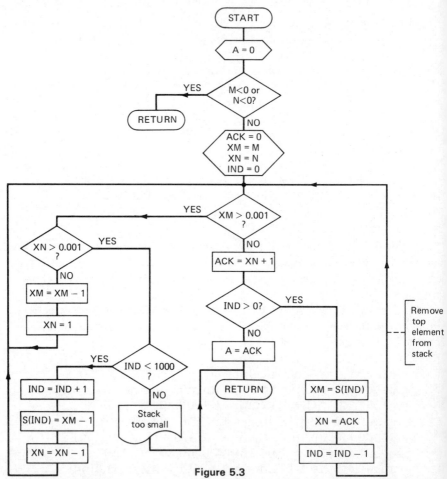

Figure 5.3

Note that we limit the stack size to 1000 elements simply to conserve storage. Also
the calculation is carried out in a real arithmetic to prevent integer overflow. After
initialization, the value of M is tested. If this is zero, and the stack is non-empty, we
continue to build up the stack. If M > 0 we move to the left-hand branch, which resets
the stack content. Let us run through the workings of the flowchart for the case A(2,2)
and compare it with the values obtained above:

Value of M = XM	Value of N = XN	Number of arguments in stack = IND	ACK	S(IND)	A(S(1),A(S(2,A(S(3), . . . S(IND),A(XM,XN) . . .)))
2	2	0	0	–	A(2,2)
2	1	1	0	S(1) = 1	A(1,A(2,1))
2	0	2	0	S(2) = 1	A(1,A(1,A(2,0)))
1	1	2	0		A(1,A(1,A(1,1)))
1	0	3	0	S(3) = 0	A(1,A(1,A(0,A(1,0))))
0	1	3	2		A(1,A(1,A(0,A(0,1))))
0	2	2	3		A(1,A(1,A(0,2)))
1	3	1	3		A(1,A(1,3))
1	2	2	3	S(2) = 0	A(1,A(0,A(1,2)))
1	1	3	3	S(3) = 0	A(1,A(0,A(0,A(1,1))))
1	0	4	3	S(4) = 0	A(1,A(0,A(0,A(0,A(1,0)))))
0	1	4	3		A(1,A(0,A(0,A(0,A(0,1)))))
0	2	3	2		A(1,A(0,A(0,A(0,2))))
0	3	2	3		A(1,A(0,A(0,3)))
0	4	1	4		A(1,A(0,4))
1	5	0	5		A(1,5)
1	4	1	5	S(1) = 0	A(0,A(1,4))
1	3	2	5	S(2) = 0	A(0,A(0,A(1,3)))
1	2	3	5	S(3) = 0	A(0,A(0,A(0,A(1,2))))
1	1	4	5	S(4) = 0	A(0,A(0,A(0,A(0,A(1,1)))))
1	0	5	5	S(5) = 0	A(0,A(0,A(0,A(0,A(0,A(1,0))))))
0	1	5	5		A(0,A(0,A(0,A(0,A(0,A(0,1))))))
0	2	4	2		A(0,A(0,A(0,A(0,A(0,2)))))
0	3	3	3		A(0,A(0,A(0,A(0,3))))
0	4	2	4		A(0,A(0,A(0,4)))
0	5	1	5		A(0,A(0,5))
0	6	0	6		A(0,6)
			7		

Note that the stack holds the current list of arguments, nested to depth IND. When XM becomes zero so that the innermost pair of arguments are of the form $A(0,p)$ we can apply the terminating condition $A(0,p) = p+1$ and store the value of $p+1$ in ACK. Then ACK will finally contain our answer when the stack is empty and XM = 0.

The FORTRAN version of this program is a straightforward translation of this flowchart, and is given below:

```
FUNCTION ACKER(M,N)
DIMENSION S(1000)
ACKER = 0.0
IF (M.LT.0.OR.N.LT.0) RETURN
ACK = 0
XM = M
XN = N
IND = 0
```
(cont.)

```
C     TEST IF M = 0
    7 IF (XM.GT.0.001) GO TO 3
      ACK = XN+1.0
C     TEST IF STACK IS EMPTY
      IF (IND.GT.0) GO TO 4
      ACKER = ACK
      RETURN
C     TEST IF N = 0
    3 IF (XN.GT.0.001) GO TO 5
      XM = XM-1.0
      XN = 1.0
      GO TO 7
    5 IF (IND.LT.1000) GO TO 6
      WRITE (6,400)
  400 FORMAT ('0 STACK CAPACITY EXCEEDED, TRY AGAIN')
      ACKER = 0.0
      RETURN
C     HERE IF M AND N ≠ 0. BUILD UP STACK
    6 IND = IND+1
      S(IND) = XM-1.0
      XN = XN-1.0
      GO TO 7
    4 XM = S(IND)
      XN = ACK
      IND = IND-1
      GO TO 7
      END
```

This program, apart from minor corrections, is due to Morris(1969), who also gives several other examples of recursive function calculation in FORTRAN. An alternative approach to programming recursive functions in FORTRAN is given by Day (1972). These two examples do, however, illustrate the point that it is quite possible to progra recursive functions in FORTRAN, though it is undoubtedly easier to do so in ALGOL Further examples are given as exercises at the end of the chapter, and in Guttmann (1976).

5.4   ALGOL and FORTRAN Compilers

While it is true that criticisms of a language must be separated from criticisms of a compiler, certain language features do place restrictions on compilers. For example, in FORTRAN the rule that spaces are ignored outside Hollerith strings means that the compiler cannot distinguish the command

$$DO5J = 1$$

ich assigns to the variable DO5J the value 1, from the command

$$DO\ 5\ J = 1,10$$

ich sets up a DO loop, until the comma in the second statement is reached. If the
)RTRAN language were modified to include a rule that spaces are not allowed in
riable names, then it would enable faster FORTRAN compilers to be written. This
oblem does not exist in ALGOL since word-like characters are written differently to
nple variables, and all variables are declared at the head of the relevant block.

In ALGOL, procedures may be recursive, yet at no stage need they be explicitly
fined as such. This means that the compiler must provide code that will allow all
ocedures to be recursive. Such object code is inefficient for non-recursive procedure
lls. Therefore, if it were necessary in ALGOL to explicitly specify that a procedure
ll is recursive, it would be possible to write a faster compiler[+].

It is the author's experience that for most straightforward, non-recursive, scientific
oblems, a FORTRAN program can be written faster, compiled faster, and debugged
ster than the corresponding ALGOL program. This may, of course, simply be a
easure of the author's relative skills in the two languages, or of the compilers he has
id to work with.

The bibliography for this chapter (see page 207) lists a number of articles which
scuss in more detail some of the points raised here.

roblems

1   The product of two positive integers n and m can be defined recursively as

$$Prod(1,m) = m$$
$$Prod(n,m) = n+m-1+Prod(n-1,m-1).$$

Write a flowchart and FORTRAN program to implement this definition.

2   There are $Q(k,k)$ different ways a positive integer k can be expressed as the sum
of non-negative integers. Thus $Q(3,3) = 3$, since $3 = 0+3 = 2+1 = 1+1+1$. $Q(k,k)$
can be defined recursively as

$$Q(n,m) = \begin{cases} 1 \text{ if } n = 1 \text{ and } m > 1 \text{ or if } m = 0 \text{ or if } m = 1 \\ 0 \text{ for } m < 0 \\ Q(n-1,m)+Q(n,m-n) \text{ otherwise.} \end{cases}$$

Write a flowchart and a FORTRAN program to implement this definition, and
test it with $Q(1,1)$, $Q(2,2)$, $Q(3,3)$, $Q(4,4)$, and $Q(5,5)$, for which you should
obtain the answers 1,2,3,5, and 7 respectively.

[+] few ALGOL compilers do check to see if any procedure can be recursive, and provide
fferent object code for recursive and non-recursive procedures.

5.3   The Towers of Hanoi problem is a fine example of a problem that can easily be
solved recursively. Given three pins, labelled 1,2, and 3 and a stack of N discs o
decreasing diameter on the first pin, the problem is to transfer these N discs to
the second pin subject to the following rules:

(1)   To move a disc means to place it on another pin.
(2)   Only one disc at a time may be moved.
(3)   At no stage may a disc be placed on top of a smaller disc.

The intial configuration given five discs is shown in Figure 5.4.

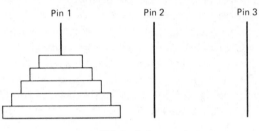

Figure 5.4

Experiment with the problem for $N = 1,2,3$. You will quickly discover the
following recursive algorithm for solving the problem:

(1)   If $N=1$, move the disc from pin 1 to pin 2 and stop.
(2)   Otherwise, move the topmost $N-1$ discs from pin 1 to pin 3.
(3)   Move the remaining disc from pin 1 to pin 2.
(4)   Move the $N-1$ discs on pin 3 to pin 2.

The recursive nature of the algorithm is clear from moves 2 and 4.

(a)   Write a FORTRAN program to implement this algorithm. Given N, the
number of discs, your program should print out the state of the pins afte
each move. Try your program for $N = 1,2,3,4,5$, and 6.
(b)   How many moves are required for N discs?
(c)   Legend has it that the original Tower has 64 discs, and that when the
problem has been completed, the universe will come to an end. Assuming
that the legend is true, and that you are solving the problem on a compu
which takes 50 ns per move, are you worried?

A detailed discussion of the algorithm is given in *A Collection of Programming
Problems and Techniques* by H. A. Maurer and M. R. Williams (New Jersey:
Prentice-Hall, 1972).

5.4   A popular activity among some computer scientists is to write a piece of softwa
called a 'preprocessor' that takes as its input a program written in a superset of
some existing language — that is, the superset contains language features not

accepted by a standard compiler — and as its output gives an equivalent program using only standard language features. For example, a preprocessor for a superset of FORTRAN which allowed recursive subroutine calls could be written.

Having studied both FORTRAN and ALGOL, consider which language features you would like to see in FORTRAN in order to be able to write better structured programs.

Write a preprocessor in FORTRAN implementing just one of these features (or more if you like). Then write a few test programs including the new language feature you have introduced, and try them out on your preprocessor.

6. Programming Style and Practice

6.1 Introduction

In this chapter we will be looking at some aspects of the general problem of program writing. Unfortunately we cannot give a strict set of rules that will lead to a perfect program, but we can give guidelines for the design and construction of good program These guidelines include the sequence of steps to be followed in program design and construction. Another aspect of program writing is the way in which certain operatic should be coded. Familiarity with these standard situations will usually lead to more efficient programs in terms of programmer's time, computer time, and memory requirements.

As discussed in Chapter 1, the recommended philosophy in program design is the top-down approach, whereby the solution to the problem is developed in a sequence steps, each of successively finer detail. By 'the solution' we mean all aspects of the solution, including the way in which the data is to be presented, the algorithm for solving the problem in the light of this form of data, and the details of the output fr the program. Since these are interdependent, the only way a solution can be efficien arrived at is by progressive refinement of each aspect. This philosophy is really just t application of the scientific method to the task of program writing.

Let us consider the creative and mechanical steps in writing a program, and then l at them in more detail. This then is a top-down approach to developing a top-down approach to program writing!

After a little reflection, it should be clear that the following six steps are involved designing a program.

(1) Clearly formulate the problem to be solved.
(2) Decide on the appropriate programming language.
(3) Decide on the best method of solution, developing your solution by flowchart by drawing decision tables, or by natural language formulation or a mixture of these representations.

4) Translate the result of part (3) into the desired programming language.
5) Test and correct the program.
6) Add final documentation to the working program.

This process is summarized in the flowchart (Figure 6.1).

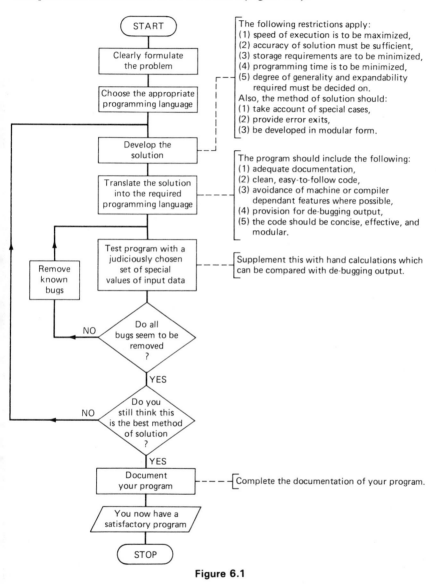

The following restrictions apply:
(1) speed of execution is to be maximized,
(2) accuracy of solution must be sufficient,
(3) storage requirements are to be minimized,
(4) programming time is to be minimized,
(5) degree of generality and expandability required must be decided on.
Also, the method of solution should:
(1) take account of special cases,
(2) provide error exits,
(3) be developed in modular form.

The program should include the following:
(1) adequate documentation,
(2) clean, easy-to-follow code,
(3) avoidance of machine or compiler dependant features where possible,
(4) provision for de-bugging output,
(5) the code should be concise, effective, and modular.

Supplement this with hand calculations which can be compared with de-bugging output.

Complete the documentation of your program.

Figure 6.1

(We will not consider the purely mechanical task of getting the program onto the desired input medium, such as cards or paper tape.) Each of these steps can be broken down into finer steps, and at each stage a number of factors have to be borne in mind. A breakdown of these steps is shown in the previous flowchart, and we will consider eac step in turn.

6.2 Formulating the Problem

This apparently obvious requirement is ignored in many cases, with programmers starting to write code with only the vaguest notion of the problem they are trying to solve. The generality of the problem must also be considered at this stage. For exampl in writing a routine to calculate the natural logarithm of x, it must be decided whethe x is to be restricted to the positive real line, the whole real line, excluding or including the origin, or the entire complex plane.

Once a clear understanding of the problem has been achieved, it is necessary to decide on the appropriate language.

6.3 Choosing a Language

This is often straightforward, particularly if the programmer only knows one language Less facetiously, for some problems a certain class of language may suggest itself. For example if a problem can be thought of recursively in a natural way, then ALGOL wil very likely be a more suitable language than FORTRAN. A problem involving comple string or list processing activities is best written in a language that specifically includes facilities for handling such data structures, such as Snobol or Lisp.

A straightforward numerical analysis problem may be most efficiently encoded in FORTRAN or BASIC. Often the appropriate language will suggest itself. If it doesn't the choice is probably not too critical, and the general-purpose language with which t programmer is most familiar is the appropriate choice.

6.4 Developing the Solution

This stage is by no means clear cut, as there are a number of conflicting requirements. These include the requirement that the execution speed of the program be as great as possible, yet storage is to be minimized. The method must be capable of giving the desired accuracy; for example, if a numerical problem is being tackled, it is well know that some numerical methods are intrinsically more accurate than others. Even this st of the process may be machine dependent to some extent. All or part of the calculatic may have to be carried out in extended-precision arithmetic if one method is used on particular machine. Single precision may suffice for the same method on another machine with a longer word length. Alternatively, some other method may permit sin precision variables to be used on the original machine. All these considerations must b taken into account at this stage.

The programmer must also consider the future fate of the program. Will it be gener alized at a later stage? Will it be expanded to cope with a much wider range of situatic If the answer to these questions is *yes*, then a method of solution must be chosen that allows for this future expansion. For example, if a program is written to find the root of a quadratic and a cubic equation, it would be sensible to program the known expre on for the roots. On the other hand, if it is likely that, at a later stage, the roots of igher-degree polynomials are required, it might be better to build in an iterative cheme to find the roots.

As a further constraint, it is clear that programmer time is to be a minimum. A short rogram that is only used once should be written more quickly than a longer program hat is to be in regular use. Thus the programmer should not spend days or weeks in the rst case trying to find the very best method of solution, and similarly, should not just ash off a method of solution in the second case without a great deal of thought.

To achieve a 'best' method of solution is clearly an optimization problem of con- iderable complexity to which no unique solution can generally be found. It is for this eason that one can consider programming an art — particularly at this stage of deve- oping the solution.

With decreasing hardware costs and increasing software costs, the relative weights iven to the different aspects of the minimization problem are changing. It is no longer sually necessary to save every last word of core storage or every microsecond of execu- on time. It is more important to have a readable program — which may involve less han optimal coding (from the machine's viewpoint) in some parts of a program. This oes not mean that core usage and execution time are unimportant, but rather that the elative cost of various resources should be judiciously balanced.

The design of the solution will involve a natural language formulation and/or the use f flowcharts or decision tables. The top-down approach is used in developing the solu- ion, and it will greatly facilitate both the design and translation of the program if the nly control structures used are the six introduced in Chapter 1.

In developing the solution it is necessary to include provision for special cases. For xample, in solving the quadratic equation $ax^2 + bx + c = 0$, provision should be made or the case $b^2 \gg 4ac$. If this is ignored, it is likely that as a result of the finite word ength of a computer, $(b^2 - 4ac)^{\frac{1}{2}}$ will yield the same value as b. In that event, one of e solutions of the equation will be given, incorrectly, as zero.

The solution should also provide for error exits if at some point in the program a ondition arises which is inconsistent with the solution of the problem. For example, a program is written to calculate the statistics of a number of examination results for class of students, it is obviously an error condition if a student has a negative mark in me subject (or possibly we have a particularly tough examiner). Another example is a rogram to solve a system of simultaneous equations. If the determinant of the co- fficients is zero, no unique solution exists. Note that in this case the determinant ay not be identically zero, but only zero to machine accuracy. In this case a nique solution may or may not exist, but an error message of some kind is still andatory. Decision sections should be carefully checked to ensure that all ranches are catered for. For example, if there is a three-way branch for $X < 0$, $X = 0$, d $X > 0$, checking is necessary to ensure that the flow from all three branches is counted for. Depending on the problem at hand, the program may be required to ntinue or abandon the job following an error exit.

If the solution has been well thought out, it will usually be seen to consist of a quence of fairly-short sections of code, each of which does a specific task. A well esigned program is built up of such *modules,* and care must be taken at the design stage to ensure that the interface between these modules is well thought out and well documented. In the translation stage, care should be taken to retain the modular natu of the program, as this will greatly facilitate program readability and maintainability.

The next phase in program development is translation of the solution into the required programming language.

6.5　Coding the Solution

In this section our remarks are principally oriented towards efficient coding of FORT programs, with less emphasis on ALGOL, BASIC, and other languages.

In translating from the flowchart into the required programming language, a numbe of points should be observed. These include:

(1)　Adequate Documentation

The importance of adequate documentation cannot be over-emphasized. If a program to receive regular use, it is likely that some modification will eventually be necessary. This is difficult, even for the original author, and virtually impossible for any subsequ user, unless there is sufficient information available about the program.

This should include such essential information as the authorship of the program, th date of birth — of the program, not the programmer — and a statement as to the function of the program. To be more specific, if the program is written in FORTRAN or ALGOL, then *comment* cards should be inserted throughout the program, explaining the purpose and function of various stages. In COBOL a whole segment of the progra the Identification Division, is set aside for this purpose. An ALGOL procedure can be identified by using the appropriate procedure delimiters — as discussed on page 95. A FORTRAN program should also state which subprograms are required, and each subprogram should be documented with its function, plus an example of the calling statement and an explanation of the various parameters. In BASIC, documenta tion is usually less important, since the programs are usually shorter. Some documenta tion is still usually necessary, however, and this can be provided by the REM statement.

If a FORTRAN program is of substantial length, say more than 100—200 cards, it i wise to use columns 73—80 to identify the main program or subprogram with a name and sequence number. For example, if the main program comprised 100 cards, those could be identified by MAIN0010 to MAIN1000. The fact that the last digit is zero allows for the insertion of extra cards at a later date, which can take numbers from, fc example, MAIN0081 to MAIN0087 if we insert seven new cards between cards 8 and in the main program. Cards can also be numbered in ALGOL using the **comment** feature.

At the head of a program it is often useful to provide a comment block giving a list of variable names and their meaning. Similarly, variable names should be chosen to be formative. For example, a FORTRAN or ALGOL statement to calculate the area of a circle could be written

$$A = B*C*C \text{ (FORTRAN)} \qquad A := bxcxc \text{ (ALGOL)}$$

$$AREA = PI*RADIUS*RADIUS \qquad AREA := pixradiusxradius$$

where the second form is, of course, vastly preferable.

In a FORTRAN program, the predefined type convention for variable names should be used. It is also a good idea to prefix logical variables with the letter L or B (for boolean), complex variables with the letter C and double precision variables with the letter D. A similar convention can also be used in ALGOL and other programming languages, even though there is no predefined type convention in many languages.

The layout of a program on cards is a valuable aid to documentation. In ALGOL the **begin** and **end** characters can be indented two or three spaces, as well as indenting the statements between them, while in FORTRAN the same applies to nested DO loops. For example,

```
begin for i:=1 step 1 until 20 do              DO 100 I = 1,20
  begin for j:=1 step 1 until N do                DO 200 J = 1,N
    begin sum:=0.0;                                 SUM = 0.0
    for k:=1 step 1 until 5 do                       DO 300 K = 1,5
    begin term:=c[i,j,k]↑j;                            TERM = C(I,J,K)**J
      sum:=sum+term                                    SUM = SUM+TERM
    end;                                    300         CONTINUE
    d[j] := sum+f[i]                                  D(J) = SUM+F(I)
  end;                                      200       CONTINUE
  ans[i] := q[i]−sum                                 ANS(I) = Q(I)−SUM
end;                                        100   CONTINUE
  ⋮                                                   ⋮
```

The program output should include headings so that the meaning of the output is clear. There are few things so frustrating as looking at a page full of numbers with no statement as to their meaning. For all but the simplest program, some documentation should be written on paper and is to be read in conjunction with the documentation within the program. This is called *external documentation,* and should include information about the minimum hardware configuration required to run the job, a statement as to which tapes or disks need to be loaded, an explanation of error fail conditions, details of any particular advanced algorithm used, full details of any amendments to the program, plus any other details that may be relevant in a particular case.

) Write Clean, Easy-to-follow Code

The program should be easy to follow and should use the minimum of obscure sections of coding. Tricky clever-clever code is to be avoided. Even if the programmer avoids the trap of outsmarting himself, he will still almost certainly outsmart the next person who looks at the program.

(3) Compiler Independence

Unless there is a very good reason, it is usually a mistake to use special features of a language provided by one particular compiler. It means the program is tied to that compiler, it cannot be run on a different machine, or even on the same machine with different compiler.

(4) Provide for Debugging Output

A large program should include easily identifiable — and thus removable — pieces of code that will print out intermediate results. These can be compared to the results of hand calculations and the sections of code giving these outputs can be removed once the program is working correctly.

The debugging output should also test the *flow* of the program. That is, it should test that the various pieces of code are being executed in the required sequence. This can most simply be done by printing out an appropriate message on entry to each block or subroutine.

A good example of programmer supplied debugging code is given by Blatt (1967).

The comments in the previous section against using compiler-dependent features also apply to manufacturer-supplied debugging packages. These seldom achieve anything that the programmer cannot achieve by including his own routines.

It also means that the programmer does not have to learn how to use a package that becomes obsolete if compilers, computers or jobs are changed.

(5) Write Short and Effective Code

Some manufacturers provide optimizing compilers which, at the expense of a substantial amount of compilation time, provide a more efficient object program. It should be the aim of the programmer to write code that can be improved as little as possible by such compilers. There is a view held in some quarters that the programmer should write code whichever way he wants to. The justification of this approach is that we are to regard the machine as the servant rather than the master. However, once one forms the habit of writing efficient code rather than inefficient code, it is no more difficult or time consuming to do the former.

There are a number of common situations for which the 'best' — that is, the most efficient — way to write a piece of code is known. In this section we will consider several such examples.

One of the best known examples is Horner's method for evaluating a polynomial. Consider the polynomial

$$D(X) = \sum_{i=0}^{n} A(i)X^i$$

ssuming that the coefficients $A(i)$, $i = 1, 2, \ldots, n$, are stored in an array A, a FORTRAN rogram to evaluate this polynomial at $X = XC$ say, may be written as follows (where 0 is assumed to be the constant coefficient of the polynomial):

```
        SUM = A0
        DO 100 I = 1,N
        SUM = SUM+A(I)*XC**I
100  CONTINUE
```

his involves $N+1$ assignment statements, N additions, N multiplications, and N expo-entiations. Horner's method expresses the polynomial as

$$(\ldots (((A(n)X+A(n-1))X+A(n-2))X+ \ldots)X+A(0)$$

hich can be programmed as

FORTRAN	*ALGOL*

```
        SUM = 0.0              sum:=a[n] ;
        NN = N+1               for i:=n-1 step -1 until 0 do
        DO 100 I = 1,N         sum:=sumxxc+a[i]
        SUM = (SUM+A(NN-I))*XC      .
100  CONTINUE                       :
        SUM = SUM+A0

        :
```

ote that in the ALGOL program we have stored the constant term in array element 0). This method of programming uses $N+3$ assignment statements (in FORTRAN — nly $N+1$ in ALGOL), $N+1$ real variable additions (N in ALGOL), $N+1$ integer variable dditions (0 in ALGOL), and N multiplications. Since exponentiation is far slower than ny of the other operations in the program, this second form of coding is clearly much ster than the first.

A more obvious example of efficient coding, but one that is frequently overlooked, the rule that 'no statement should be within the range of a loop that is invariant roughout the loop'. For example, the loop

```
        DO 100 I = 1,100
        PI = 3.141592654
        A(I) = PI*R*R/FLOAT(I)
100  CONTINUE
```

should be written

$$PI = 3.141592654$$
$$PIRSQ = PI*R*R$$
$$DO\ 100\ I = 1,100$$
$$A(I) = PIRSQ/FLOAT(I)$$
$$100\ \ CONTINUE$$

thus saving 99 executions of the statements PI = 3.141592654 and PI*R*R. The previous rule is a special case of the more general rule 'do not evaluate anything more oft than necessary'. For example, to calculate the volume of a sphere, area of a sphere, an area of a circle of radius r, we have

$$v = \tfrac{4}{3}\pi r^3,\ sa = 4\pi r^2,\ a = \pi r^2,$$

respectively, which might be programmed

$$PI = 3.141592654$$
$$VOL = 4.0*PI*R**3/3.0$$
$$AREAS = 4.0*PI*R**2$$
$$AREAC = PI*R**2$$

A much better way to program this is the following

$$PI = 3.141592654$$
$$AREAC = PI*R*R$$
$$AREAS = 4.0*AREAC$$
$$VOL = AREAS*R/3.0$$

In the first case there are 13 multiplications and divisions (assuming that the exponer tiations are carried out by repeated multiplication), while in the second case there are only 5 multiplications and divisions.

The rule that 'the number of function calls should be minimized' is obvious enoug yet it is possible to make a function call without noticing that one is being made. For example, it is well known that

$$A = B**2$$

is executed much more rapidly than

$$A = B**2.0$$

since the former is usually translated

$$A = B*B$$

while the latter is equivalent to

$$A = EXP(2.0*ALOG(B))$$

which involves two (slow) out-of-line function calls.

Less obvious, perhaps, is that

$$I = A$$
$$J = K+I$$
$$L = M+I$$

is more efficient than

$$J = K+A$$
$$L = M+A$$

since the second form required two calls to IFIX, while the first form requires only one. In this example a slight increase in speed is gained at the cost of a decrease in readability, so this practice should only be used if speed is of utmost importance.

Subscript manipulation is often a very efficient way of cutting down execution time. Most computers store arrays in a linear sequence of addresses so that the two-dimensional array $A(5,5)$ uses 25 adjacent words in memory. The first word corresponds to $A(1,1)$, the second word to $A(2,1)$, the third word to $A(3,1)$, and so on. In multi-dimensional arrays, the *mapping* between the array elements and the memory locations is such that successive memory locations correspond to the leftmost subscript changing most quickly and the rightmost subscript changing least quickly. Thus, the three-dimensional array $A(10,10,10)$ is stored in 1000 successive words, with element $A(4,5,6)$ corresponding to the $4+(5-1) \times 10^1 +(6-1) \times 10^2 = 544$th memory location. Each call to a multi-dimensional array element requires a calculation by the computer to transform the array subscript into the machine address. Thus, the higher the dimension of the array, the more time-consuming is a call to a particular element. Therefore, if different multi-dimensional arrays use the same subscripts, it is a considerable time-saver to calculate the *mapping function* explicitly. Consider the following program segment:

```
DIMENSION A(5,5,5),B(5,5,5),C(5,5,5)
  .
  .
  .
X = A(I,J,K)+2.0*B(I,J,K)+3.0*C(I,J,K)
  .
  .
```

This can be more efficiently coded as

```
DIMENSION A(5,5,5),B(5,5,5),C(5,5,5),AA(125),BB(125),CC(125)
EQUIVALENCE (A(1,1,1),AA(1)),(B(1,1,1),BB(1)),(C(1,1,1),CC(1))
  .
  .
  .
C    THIS IS THE MAPPING FUNCTION BETWEEN A(I,J,K) AND AA(L)
L = 25*(K-1)+5*(J-1)+I
X = AA(L)+2.0**B(L)+3.0*CC(L)
  .
  .
```

A good compiler would do all this automatically, but there are many compilers around that would not. To clear even one of these arrays to zero, it is more efficient to write

DO 100 I = 1,125		DO 100 I = 1,5
AA(I) = 0.0	than	DO 100 J = 1,5
100 CONTINUE		DO 100 K = 1,5
		A(I,J,K) = 0.0
		100 CONTINUE

Note that the principle discussed here applies to languages other than FORTRAN, but the mapping function is likely to vary from one language to another.

There are a number of ways to shorten the execution times of certain types of loop. Here are some of them.

If a test takes place inside a loop that is independent of the loop variable, it should be removed from the loop if possible. For example

```
DO 100 I = 1,100
IF(TEST.LT.TOLER)GO TO 50
A(I) = B(I)+C(I)**2
GO TO 100
50 A(I) = C(I)+B(I)**2
100  CONTINUE
```

can be more efficiently coded as

```
IF(TEST.LT.TOLER)GO TO 50
DO 100 I = 1,100
A(I) = B(I)+C(I)**2
100  CONTINUE
GO TO 200
50 DO 150 I = 1,100
A(I) = C(I)+B(I)**2
150  CONTINUE
200  CONTINUE
```

Though this second program segment is longer, it will execute much more rapidly since the IF statement is evaluated only once. It is also more readable than the first example. The corresponding ALGOL example is more readable still, as you can convince yourself by writing it out.

Since addition is usually a faster operation than multiplication, it is often possible to save time by replacing a multiplication with an addition; for example

DO 100 I = 1,100	can be replaced by	J = −2
J = 4*I−2	the more efficient	DO 100 I = 1,100
	and more readable	J = J+4
	program segment	
.		.
:		:
.		.
100 CONTINUE		100 CONTINUE

which saves one multiplication for each loop pass. Note that if the loop variable I were not used throughout the loop, an even faster version would be

```
        DO 100 J = 2,398,4
                :
                :
100  CONTINUE
```

f two loops have the same limit, it is often possible to combine the loops (this is known as 'jamming'). For example,

```
        DO 100 I = 1,100
        A(I) = X**I
100  CONTINUE
        DO 200 I = 1,100
        B(I) = A(I)+X
200  CONTINUE
```

an be replaced by the more efficient and more readable single loop

```
        DO 100 I = 1,100
        A(I) = X**I
        B(I) = A(I)+X
100  CONTINUE
```

mplied DO loops in input/output statements should be avoided whenever possible. Thus, if array VEC has dimension 200, the statement WRITE(3,100) (VEC(I),I=1,200) s very much slower than WRITE(3,100)VEC. For efficient data initialization, BLOCK DATA and DATA statements should be used. For example

```
        DIMENSION A(20), WEIGHT(4)
        DO 10 I = 1,20
        A(I) = 0.0
10   CONTINUE
        PI = 3.1415927
        WEIGHT(1) = 1.0
        WEIGHT(2) = 4.0
        WEIGHT(3) = 2.0
        WEIGHT(4) = 1.0
```

hould be replaced by

```
        DIMENSION A(20), WEIGHT(4)
        DATA A/20*0.0/,WEIGHT/1.0,4.0,2.0,1.0/,PI/3.1415927/
```

This saves both storage space and execution time. If it is required to have these arrays accessible to all subprograms, the arrays could be put in labelled COMMON, which can then be initialized in a BLOCK DATA subprogram. For example, to put the above array WEIGHT(4) and the constant PI in COMMON, we would use the BLOCK DATA subprogram

```
BLOCK DATA
COMMON/JUNK/WEIGHT(4),PI
DATA WEIGHT/1.0,4.0,2.0,1.0/,PI/3.1415927/
END
```

The statement

```
COMMON/JUNK/WEIGHT(4),PI
```

must, of course, appear in any program unit (main program or subprogram) which may use of the variables initialized in the BLOCK DATA statement.

If a parameter passed in a subroutine call is used frequently in that subroutine, it is often more efficient to either: (a) re-set the parameter in the subroutine, or (b) pass the parameter by a COMMON statement. For example

```
SUBROUTINE EXAMPL(X,Y,SUM,PROD,QUOT,DIFF)
SUM = X+Y
PROD = X*Y
QUOT = X/Y
DIFF = X−Y
    .
    .
    .
```

is more efficiently written as

```
SUBROUTINE EXAMPL(X,Y,SUM,PROD,QUOT,DIFF)
XLOCAL = X
YLOCAL = Y
SUM = XLOCAL+YLOCAL
PROD = XLOCAL*YLOCAL
QUOT = XLOCAL/YLOCAL
DIFF = XLOCAL−YLOCAL
    .
    .
    .
```

since in the first example every time X or Y is referred to, their address, as passed by the subprogram call, must be inserted into the instructions which fetch the values of the dummy parameters. Re-assigning the dummy variables to local variables, as in the second program segment, overcomes this problem.

A still more efficient and more readable way to write this subroutine is

```
SUBROUTINE EXAMPL
COMMON/SUB1/X,Y,SUM,PROD,QUOT,DIFF
SUM = X+Y
PROD = X*Y
QUOT = X/Y
DIFF = X−Y
    ⋮
    ⋮
```

with, of course, a similar COMMON statement in the calling program.

In addition to the programming hints already discussed, another aspect of writing efficient code is the elimination of unnecessary storage space. If storage space is at a premium, the program can be written making use of *overlays*, in which the program is broken up into parts (by the programmer) which are not all in high-speed core store at the same time. As needed, these segments are called in from a lower speed peripheral device and placed in high-speed core. Once a segment is not needed, the next required segment is called in, *overlaying* the area of core occupied by the previous segment. Sometimes a copy of the no-longer needed segment is saved on backing store before overlaying, but only if it is to be used again.

Some manufacturers provide *paging* on their machines (also called *virtual memory*) in which the operating system (sometimes aided by paging hardware) 'chops up' the program into pages which are stored on peripherals, and are called in by the operating system as required. The availability of paging is by no means widespread, so the programmer usually has to do the work by overlaying his program.

A lot of storage space can often be saved by reducing the storage required by arrays, especially multi-dimensional arrays. An obvious example is the case when two arrays are to be used sequentially. That is, array A(20,20,20) is finished with before array B(100,80) is used. Then writing

```
DIMENSION A(20,20,20),B(100,80)
EQUIVALENCE (A(1,1,1),B(1,1))
```

allows the two arrays to share the same storage locations. In ALGOL this is automatically achieved by the block structure, so that in the following example

```
          begin real a,b,c, . . . ;
              ⋮
          begin real array a[1:20,1:20,1:20] ;
              ⋮
Block B1      ⋮
              ⋮
          end
```

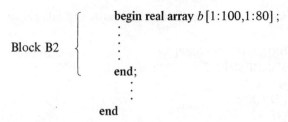

$$\text{Block B2} \begin{cases} \textbf{begin real array } b[1{:}100,1{:}80]\,; \\ \quad \vdots \\ \quad \vdots \\ \textbf{end;} \end{cases}$$

$$\vdots$$

$$\textbf{end}$$

arrays a and b have *disjoint scope* so they can share the same storage locations.

In dealing with arrays in which a number of array elements are zero, it is necessary only to store the non-zero elements. For example, a triangular square matrix A

$$A = \begin{pmatrix} a_{11} & a_{12} & & & a_{1n} \\ 0 & a_{22} & & & a_{2n} \\ 0 & 0 & & & \vdots \\ \vdots & \vdots & & & \vdots \\ 0 & 0 \ldots\ldots\ldots 0 & & & a_{nn} \end{pmatrix}$$

has all its matrix elements equal to zero below the diagonal — or above the diagonal. T store this as a square matrix requires n^2 locations, but storing only the non-zero eleme requires only $n(n+1)/2$ locations, a saving of almost 50 per cent. The array can be store in a one-dimensional array TRIANG, with dimensions $n(n+1)/2$ with TRIANG(1) = a_{11} TRIANG(2) = a_{12}, TRIANG(3) = a_{22}, TRIANG(4) = a_{13}, and so on. The only problem here is the construction of the *mapping function* between the original array and the ne array. The mapping between A and TRIANG is defined by

$$A(I,J) = TRIANG(J*(J-1)/2+1) \text{ for } J \geqslant I$$
$$A(I,J) = 0.0 \qquad\qquad\qquad \text{for } J < I$$

In the main program, TRIANG is dimensioned (and possibly placed in COMMON) whi A is not dimensioned. References to A(I,J) therefore constitute a FUNCTION call. Th A(I,J) becomes a function subprogram call to the following subprogram:

```
      FUNCTION A(I,J)
      COMMON TRIANG( ...... )
      IF (J.GE.I) GO TO 20
      A = 0.0
      RETURN
   20 NARG = J*(J-1)/2+I
      A = TRIANG(NARG)
      RETURN
      END
```

Similarly, a symmetric square matrix of N rows and columns — defined by $A(I,J) = A(J,I)$ — needs only $N(N+1)/2$ storage locations, since only the upper or lower triangular part needs to be stored. Thus, if the upper triangular part is stored as above, we can retrieve all the elements using the following mapping function:

$$A(I,J) = TRIANG(J*(J-1)/2+1) \quad \text{for } J \geqslant I$$
$$A(I,J) = A(J,I) \quad \text{for } J < I$$

Similar considerations apply to other irregularly shaped arrays; for example, a square matrix with non-zero elements only on the main diagonal and the two diagonals on either side. It is just a matter of working out the mapping function between the initial square array and the one-dimensional array in which the non-zero elements are stored.

It should be noted that this saving in storage space comes at the cost of decreased readability and increased execution time, caused by the evaluation of the mapping function at each array call. Therefore this technique should only be used when saving storage space is important.

Another common type of array storage problem is the storage of *sparse matrices*. These are matrices in which most of the matrix elements are zero and those that are not zero do not fall into any regular pattern. Consider the following 6x6 sparse matrix A.

$$A = \begin{pmatrix} 0 & 1 & 0 & 3 & 0 & 0 \\ 0 & 0 & 7 & 9 & 0 & 0 \\ 4 & 0 & 0 & 0 & 0 & 0 \\ 0 & 6 & 0 & 0 & 9 & 0 \\ 0 & 0 & 3 & 0 & 2 & 0 \\ 5 & 0 & 0 & 0 & 0 & 8 \end{pmatrix}$$

Of the 36 matrix elements, only 11 are non-zero. Storage space can be saved by storing only the non-zero matrix elements in a one-dimensional array, and storing a number that identifies the position of that element in a second array. The mapping from the one-dimensional array to the original array then requires a search through the second one-dimensional array to see if the required matrix element is listed here.

In the above example, we first consider the mapping function from A to a one-dimensional array C. The column-wise mapping from $A(I,J) \rightarrow C(K)$ is achieved by

$$K = I+(J-1)*6$$

The non-zero matrix elements of A and their corresponding location in C is shown in the table on page 132.

It is only necessary to store the two one-dimensional arrays IE and IPOS which store the 11 non-zero matrix elements and the 11 corresponding positions in array C respectively. That is,

$$IE(1) = 4, IE(2) = 5, IE(3) = 1, IE(4) = 6, \dots, IE(11) = 8 \text{ while}$$
$$IPOS(1) = 3, IPOS(2) = 6, IPOS(3) = 7, \dots, IPOS(11) = 36.$$

Element	Position in A	Position in C
4	A(3,1)	C(3)
5	A(6,1)	C(6)
1	A(1,2)	C(7)
6	A(4,2)	C(10)
7	A(2,3)	C(14)
3	A(5,3)	C(17)
3	A(1,4)	C(19)
9	A(2,4)	C(20)
9	A(4,5)	C(28)
2	A(5,5)	C(29)
8	A(6,6)	C(36)

The FUNCTION subroutine to achieve the mapping is given below, together with an example of its use with a calling program:

```
      COMMON/ARRAY/IE(11),IPOS(11)
C     NOW SET UP THE ARRAYS IE AND IPOS
      :
      :      (arrays set here)
      :
      :      (more programming here)
      :
      JCX = M+IA(I,J)
C     IA(I,J) IS A FUNCTION SUBROUTINE CALL WHICH RETURNS
C     THE REQUIRED MATRIX ELEMENT
      :
      :
      :
      END


      FUNCTION IA(I,J)
      COMMON/ARRAY/IE(11),IPOS(11)
      K = I+(J-1)*6
C     THIS MAPS FROM A 2 DIMENSIONAL ARRAY IA(I,J) TO A
C     1 DIMENSIONAL ARRAY C(K)
C     NOW SET UP A LOOP TO TEST IF THE MATRIX ELEMENT
C     IS ZERO OR NOT
      DO 100 M = 1,11
      L = M
      IF (IPOS(M)-K)100,200,300
  100 CONTINUE
```

(cont.)

```
C      WE REACH THIS POINT IF MATRIX ELEMENT IS ZERO
  300  IA = 0
       RETURN
C      WE REACH THIS POINT IF MATRIX ELEMENT IS NON-ZERO
  200  IA = IE(L)
       RETURN
       END
```

One further rule in writing clean, easy-to-follow code is to break up the program at logically reasonable points into subroutines. This has many advantages, among which are:

(1) it is easier to trace through a program which has been broken into a number of subroutines than to wade through a single program of the same length;
(2) each subprogram can be compiled independently;
(3) subprograms can be written by different people, which makes it a great deal easier to split up a large programming task among several people.

Finally, if the program runs for a long time, it is a good idea to provide *dump-restart* facilities. This may be the print-out of intermediate results every few minutes, so that the program can be restarted from the last set of results if necessary. Alternatively, a core-dump onto disc or tape may take place every few minutes. The purpose of this, of course, is to guard against system or program failure after a lot of useful calculation has been done, yet before the final answers appear.

The foregoing gives some idea as to points that should be considered when translating from a solution into a programming language. There are many areas we have not discussed, notable amongst which are the numerical problems due to the finite length accuracy to which a computer works. In certain cases the rounding errors in a problem can destroy all the significant decimal digits in a calculation. Consideration of the reduction of these errors more properly belongs to a numerical analysis course, and we will not discuss this topic further. Further discussion of these and other examples, with particular application to FORTRAN, is given by Kreitzberg and Shneiderman (1972), Larson (1971), and Kernighan and Plauger (1974).

6.6   Testing the Program

Following the writing of a program, the testing process can begin. Firstly, the punching and syntax errors are removed so that the program can compile and execute. Next, the flow is tested, as discussed in the section on debugging. Once the flow is satisfactory, intermediate results may be studied using programmer-supplied debugging facilities. These intermediate results should be compared with hand calculations of certain special cases.

The whole testing process must be repeated for several special cases. It is the duty of the programmer to think of all kinds of ways to test his program, and not just to choose one simple case. For example, if a program is written to solve the quadratic equation

$ax^2+bx+c = 0$, it is *not* sufficient to try it out with $a = 1$, $b = 2$, and $c = 1$. A number of test cases should be tried, including:

(1)   a data set with $a = 0$;
(2)   a data set with $b^2 - 4ac < 0$;
(3)   a data set with $b^2 \geqslant 4ac$.

In a long and complex program there will probably be at least one special case which the programmer neglects, but as many special cases as possible should be tested at the beginning. At this stage the programmer occasionally comes to the disheartening realiza tion that he made a mistake at the second stage of the program writing process. That is, his proposed method of solution is *not* the best possible. The proper action at this poin is not to throw-up ones hands in despair and wander off disconsolately to the nearest bar, but to start again, this time with the correct method of solution.

If the program has been designed in a top-down manner however, it is likely that all special cases will have been included, and that the program is satisfactory right from the start. Top-down design also means that it is most unlikely that the best method of solution is only arrived at when another method of solution has been completely pro- grammed. The process of stepwise refinement has as a consequence the fact that de- ficiencies in design strategy are likely to manifest themselves quite early in the process of programming.

Once any bugs revealed by the testing process have been removed, the programmer has a program which will probably be satisfactory, though if it is in frequent use, it will no doubt be modified and improved from time to time. It is therefore necessary, *at this stage*, to add the final documentation giving all pertinent details of the program — though again, a top-down approach to design includes progressive documentation, so that the amount of final documentation should be minimal.

Problem

Write the programs specified at the end of the section on BASIC in FORTRAN using th methods discussed in this chapter. You should find these are now more efficient, and at the same time more readable.

7. Selecting an Algorithm

7.1 Algorithms, Procedures, and Heuristics

In the previous chapter we studied the problem of program design, testing, and construction. In this chapter we shall be studying the most creative part of program writing, which is the design of the method of solution. For most programs that we write, this involves the selection of a sequence of calculational steps, each of which is clearly defined, in such a manner that the calculation terminates. Such a sequence of steps is dignified with the name of an algorithm. Knuth (1968) gives an interesting history of the word, and ascribes its origin to the name of the Arabic author, Abu Ja'far Mohammed ibn Mûsâ al-Kowârizmi, whose works appeared around 825 A.D.

The word was originally algorism, and was gradually changed to algorithm during the passage of the centuries. Until about 1950 it was traditionally associated with Euclid's algorithm for finding the greatest common divisor of two integers, but since that time has been used for a wide variety of processes.

There are a number of definitions of the word algorithm in the literature; for example, *The International Dictionary of Applied Mathematics* (Princeton, New Jersey: D. Van Nostrand & Co Ltd, 1960) defines an algorithm as follows: 'The term algorithm is used to denote — (1) any method of computation, whether algebraic or numerical, or (2) any method of computation consisting of steps to be taken in a pre-assigned order which are specifically adapted to the solution of a problem of some particular type'. The ANSI and Australian Standards read: 'A prescribed set of well-defined rules or *processes* for the solution of a problem in a finite number of steps; for example, a full statement of an arithmetic procedure for evaluating $\sin(x)$ to a stated precision'. The definition given by Knuth is perhaps the most appealing, and paraphrasing Knuth's definition somewhat, an algorithm is 'a recipe or process or procedure, method, technique or routine in that it is a finite set of rules which gives a sequence of operations for solving a specific type of problem. It has five important features:

1) finiteness — that is, the algorithm must terminate;

(2)　definiteness — that is, each statement must be clear, unambiguous, and definite;

(3)　input — that is, an algorithm has $n \geqslant 0$ inputs;

(4)　output — that is, an algorithm has $m \geqslant 1$ outputs;

(5)　effectiveness — that is, the operations involved must be sufficiently basic that the can be done exactly in a finite and reasonable length of time.'

These definitions can best be illustrated by example, and we choose that paradigm, Euclid's algorithm, which finds the greatest common divisor (g.c.d.) of two positive integers m and n. (Note that the g.c.d. of *any* integers m and n is equal to the g.c.d. of $|m|$ and $|n|$, so that Euclid's algorithm may be used to find the g.c.d. of any two integer Euclid's algorithm may be expressed in natural language and flowchart form as:

INPUT m and n;
REPEAT
　　　　BEGIN divide m by n, leaving remainder r, such that $0 \leqslant r < n$;
　　　　　　IF $r \neq 0$ THEN BEGIN
　　　　　　　　　　　　　$m \leftarrow n$;
　　　　　　　　　　　　　$n \leftarrow r$;
　　　　　　　END
　　END
UNTIL $r = 0$;
Print n as greatest common divisor

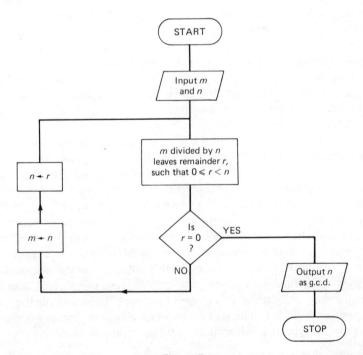

Figure 7.1

For example, the g.c.d. of 279 and 729 is calculated as follows:

$$279 = 0.729 + 279$$
$$729 = 2.279 + 171$$
$$279 = 1.171 + 108$$
$$171 = 1.108 + 63$$
$$108 = 1.63 + 45$$
$$63 = 1.45 + 18$$
$$45 = 2.18 + 9$$
$$18 = 2.9 + 0$$

$$\therefore \text{g.c.d.} = 9$$

Clearly, this algorithm satisfies all the requirements of an algorithm. Note that the algorithm need not be algebraic, and can be a process for almost anything. For example, an algorithm to fill a ditch with sand may be stated as follows:

Place a pile of sand next to the ditch. Obtain a shovel. Shovel sand into the ditch. When the ditch is full, stop. If you run out of sand, get more sand and continue shovelling. Repeat this process until the ditch is full, then stop.

The algorithm is shown below written in natural algorithmic language and in flow-chart form in Figure 7.2 on page 138.

```
Place a pile of sand next to the ditch;
Obtain a shovel;
REPEAT
        BEGIN
                IF the pile of sand is empty THEN get more sand;
                Place a shovelful of sand into the ditch
        END
UNTIL the ditch is full.
```

This algorithm satisfies all five criteria of Knuth's definition if we extend the concepts of input and output, so that *sand* is input, and a *full ditch* is the output. A purist might quibble that this is not a very good example of an algorithm in that the steps are not defined sufficiently precisely; for example, where is the shovel to be obtained from and where is the sand to come from? This demonstrates that there is a certain amount of assumed background information in any algorithm. In an algorithm for a cake (that is, a recipe) the rudiments of cooking techniques are assumed. In a mathematical algorithm, elementary operations like addition and multiplication are assumed.

Note that algorithms are not usually unique. For example, to find the g.c.d. of two positive integers there is another algorithm due to R. Silver and J. Terzian (unpublished) and independently discovered by J. Stein (1967) (*see also* Knuth (1968)) which may well be faster than Euclid's algorithm. This algorithm, called the *binary g.c.d. algorithm* is defined as shown on page 138.

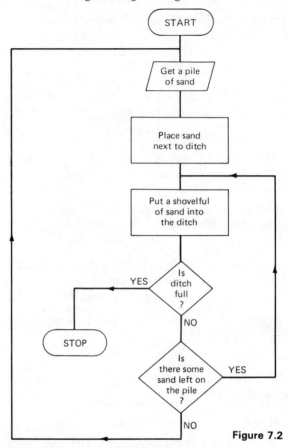

Figure 7.2

To find the g.c.d. of two positive integers m and n,
Initialize k to zero;

WHILE m and n are both even DO BEGIN

$$m \leftarrow m/2;$$
$$n \leftarrow n/2;$$
$$k \leftarrow k + 1$$

END

IF m is even THEN BEGIN

$$j \leftarrow m;$$
$$j \leftarrow j/2$$

END

ELSE $j \leftarrow m$;

WHILE $j \neq 0$ DO BEGIN

WHILE j is even DO $j \leftarrow j/2$;
IF $j > 0$ THEN $m \leftarrow j$ ELSE $n \leftarrow -j$;
$$j \leftarrow m - n$$

END

print $m.2^k$ as the greatest common divisor.

The binary g.c.d. algorithm is shown in flowchart form below:

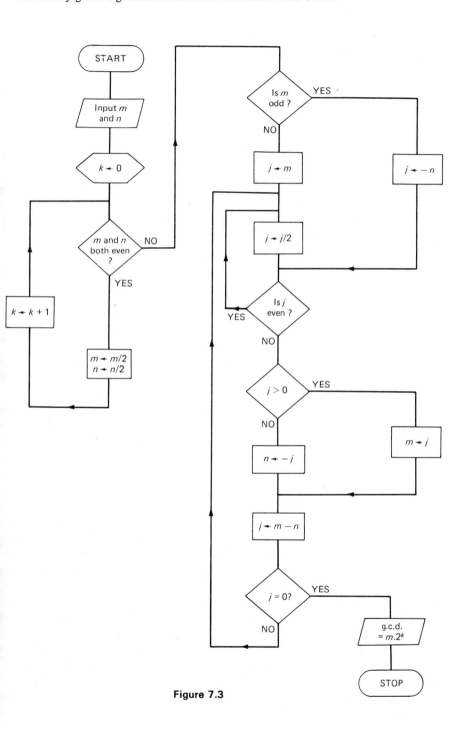

Figure 7.3

Applying this algorithm to the same pair of numbers, $m = 729, n = 279$, which were used to demonstrate Euclid's algorithm, we get the following sequence:

$$m = 729, \quad n = 279, \quad \therefore \; k = 0$$
$$j = -279, \quad n = 279.$$
$$j = 450, \quad j = 225, \quad m = 225$$
$$j = -54, \quad j = -27, \quad n = 27$$
$$j = 198, \quad j = 99, \quad m = 99$$
$$j = 72, \quad j = 36, \quad j = 18, j = 9, m = 9$$
$$j = -18 \quad j = -9 \quad n = 9$$
$$j = 0, \quad \text{g.c.d.} = 9$$

It is clear that the binary algorithm will generally involve more steps than Euclid's algorithm. However, the operations that take place at each step are considerably simple the most complicated operation being the halving of an even number. This operation is particularly suitable for implementation on an internal binary computer as it can be achieved by shifting the number 1 bit to the right; for example, to halve the *even* binary number 1011011010 we shift the number to the right one bit and thereby lose the end zero, and obtain 101101101. This shift operation cannot, of course, be programmed in a high-level language like FORTRAN, and it is therefore necessary to program in assembly language if a high speed version of this algorithm is desired.

Unfortunately, we cannot write *algorithms* for all the problems we would like to solve — either for the theoretical reason that no algorithm exists for the solution of the problem at hand, or else because the problem is of such complexity that the criterion of 'reasonable length of time' is violated. In the first case we refer to a **procedure**, which can be defined in the same way as an algorithm without the restriction that it terminat A trivial example of a procedure is a program to print out the decimal expansion of a given rational number, stopping when there are no non-zero digits in the expansion to be printed out. Clearly, the program terminates for $\frac{1}{4}$ but does not terminate for $\frac{1}{3}$. A non-trivial example of a procedure is a program that searches for a counter-example to Fermat's Last Theorem, which states that there is no solution of $x^n + y^n = z^n$ for non-zero integers $x, y,$ and z if n is an integer greater than 2. One can check for successively higher values of n, but if the program does not terminate, the theorem is neither prove nor disproved. If the program terminates, you have disproved the theorem and you are famous! The point is however that the program need not terminate. It is of interest to note that it can be proved that no algorithm exists which tests whether any given program halts. Clearly, a procedure exists, the procedure being to simply run the program in question. If it halts, then obviously it halts. If it does not halt, then we are no better off, since maybe it would halt if we let it run longer, or maybe it really doesn't halt. A remarked above, it can be proved that no algorithm for this task exists.

Some problems are simply too complex for us to undertake their complete solution An example of such a problem is that of an *optimal* chess playing program. By optima we mean a program that plays the best possible game. Clearly an algorithm exists to simply play chess — one such algorithm is: When it is your first move, resign. This is a legitimate sequence of moves within the rules of the game. It is not however, the basis of an algorithm that is likely to receive widespread acclaim. In attempting to design a program to play an optimal chess game most programs look ahead a few moves (typically less than seven), and maximize the position with respect to some basic guidelines. Thus what the program really does is to make suggestions as to moves that are likely to be reasonable – so it is a guide to a reasonable chess game, but it is not by any means an optimal solution. Such a program is sometimes called a *heuristic*.

In the remainder of this chapter we will be looking at the problem of designing and analysing algorithms, and will not be concerned with procedures and heuristics.

7.2 Design of Algorithms

When faced with a programming problem, one can either design one's own algorithm or use an algorithm designed by somebody else, or judiciously blend both approaches. For a number of numerical processes, many available algorithms have been thoroughly investigated, and their properties, such as run time, storage requirements, and accuracy well documented. For example, if one is manipulating polynomials, it is well known that Horner's method (discussed in Chapter 6) for evaluating polynomials is optimal in terms of the arithmetic operations required. Thus to design some other method of polynomial evaluation is clearly a waste of time.

When designing a program in a top-down manner, there will be a stage where the various calculational steps are sufficiently explicit for one to try to use available algorithms. For example we may reach a stage in the program design where we have to multiply two matrices together, or solve a set of simultaneous equations, or sort a file into ascending order. To program these operations it is sensible to consult the literature for details of existing algorithms and use these. On the other hand, if your program involves either very unusual calculations – so that no algorithms are likely to exist – or very specific calculations, you should design your own algorithm. In the first case you have no choice, since there is no existing algorithm for your problem. In the second case it is possible that you can improve on existing *general* algorithms. For example, there are very efficient general algorithms for multiplying two matrices together. If your specific problem involves multiplying two matrices together, with one matrix known to be diagonal, it is easy to design an algorithm that performs much faster than the fastest available *general purpose* algorithm.

It is not possible to be completely specific about designing an algorithm, since this is essentially a creative process. Guidelines to this process would include top-down design philosophy, constructing algorithms from simpler algorithms, and, for serious exponents of the art, a knowledge of how different types of data can best be structured and manipulated, and how and when to use certain fundamental techniques such as recursion and dynamic programming. These topics are beyond the scope of this book, but an introduction to some of the ideas alluded to here can be found in Knuth (1968) and Aho, Hopcroft, and Ullman (1974).

7.3 Analysis of Algorithms

As we saw in Section 7.1, the choice of algorithm for any given problem is not usually unique. If we have in mind the implementation of the algorithm on a computer, there are three features that must be compared in deciding on one algorithm rather than another. These are

(1)   storage requirements,
(2)   accuracy of the answer, and
(3)   time taken (which depends on the number of times each step is executed).

The storage requirements of a particular algorithm can usually be estimated fairly accurately. For example, the storage requirements of Euclid's algorithm comprise storage for three integer variables, plus storage for the half dozen or so instructions that comprise the algorithm.

The accuracy of the answer given by an algorithm depends firstly on whether *integer* or *real* arithmetic is used. If integer arithmetic is used, we can expect the answer to be exact, provided that all calculations lie within the word size of the computer. For example, Euclid's algorithm gives an exact, integral answer. If real arithmetic is used, the accuracy of any algorithm is limited by rounding errors. The effect of these may vary from totally unimportant to completely disastrous. The study of propagation of rounding errors is, however, more properly a subject for study in a numerical mathematics course, so we will not discuss it further at this stage – which should not be taken to imply that it is unimportant.

The time taken for a particular algorithm is estimated by multiplying the time taken for each step by the number of times the step is performed. The time taken for each step varies from machine to machine, but can usually be readily obtained from the manufacturer's data. Let us look at some examples.

Example 1
To evaluate the polynomial $a_n x^n + a_{n-1} x^{n-1} + \ldots + a_1 x + a_0$ from the above expression clearly requires n additions, n multiplications, and $n-1$ exponentiations. Writing the polynomial in the form used by Horner's method, $(\ldots ((a_n x + a_{n-1}) x + a_{(n-2)}) x \ldots \ldots + a_1) x + a_0$, we see, by inspection, that n multiplications and additions are needed, but that no exponentiations are used. Clearly, Horner's method is greatly superior.

Example 2
To multiply two $n \times n$ matrices together, that is $A \times B = C$, the coefficient

$c_{ij} = \sum_{k=1}^{n} a_{ik} b_{kj}$. There are thus n multiplications and $(n-1)$ additions involved in evaluation c_{ij}, so that there are n^3 multiplications and $n^3 - n^2$ additions involved in multiplying two such matrices, since there are n^2 matrix elements in C. Any other proposed method of matrix multiplication must be evaluated in this way, and should only be used if it is faster, or has some other advantage.

Example 3
Given a list of n distinct integers in random order, $X(1), X(2), \ldots, X(n)$, find the

ntegers m and j such that $m = X(j) = \max_{1 \leqslant k \leqslant n} \{X(k)\}$. That is, m is the maximum integer, nd j indexes its position in the list.

The algorithm used is quite straightforward and can be described as follows, and with the flowchart shown in Figure 7.4 overleaf.

Set $j \leftarrow n$; $k \leftarrow n{-}1$; $m \leftarrow X(n)$;
WHILE $k{\neq}0$ DO BEGIN
          IF $X(k) > m$ THEN BEGIN
                            $j \leftarrow k$;
                            $m \leftarrow X(k)$
                   END
                  $k \leftarrow k{-}1$;
   END
print m,j

Note that the algorithm handles the list from the nth element to the first.

The storage requirements for this program are minimal. One word is required for each of $X(n)$, $X(k)$, j,k,m, and n – and even this can be reduced since, for example, $X(n)$ and m can share the same storage location. The program itself also requires some storage, but obviously not very much, being a short and simple program.

The time requirements of this algorithm depend on the frequency with which each step is executed, and this is summarized in the table below for the eight steps shown in the flowchart (Figure 7.4).

> step number 1 is performed once
> step number 2 is performed once
> step number 3 is performed n times
> step number 4 is performed once
> step number 5 is performed $n{-}1$ times
> step number 6 is performed $n{-}1$ times
> step number 7 is performed A times
> step number 8 is performed $n{-}1$ times

We therefore only need to determine A – the number of times step 7 is evaluated – to have essentially completed the analysis of this algorithm. Note that the algorithm treats the file in order $X(n) \ldots X(1)$. If $X(n)$ is the maximum element, then $m=X(n)$ and $=n$ and $A=0$. If the file is in the order $X(n) < X(n{-}1) < X(n{-}2) < \ldots \ldots < X(1)$, then $n{-}1$ interchanges will be needed to find the maximum, that is, $A = n{-}1$. Since step 7 is not a particularly time consuming one, it is sufficient for most purposes to know that $0 \leqslant A < n$. Then by multiplying the time taken for each step by the number of steps, we can calculate the time requirements. If one requires more accuracy, Knuth (1968) has shown how to carry out a detailed analysis to determine the average value

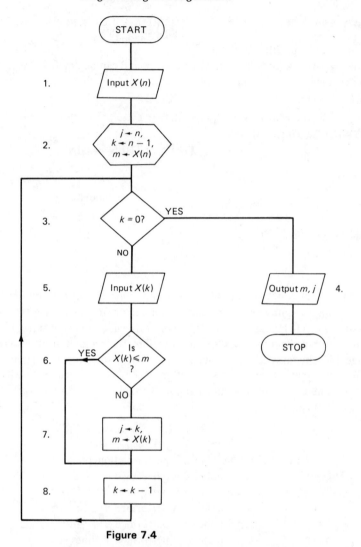

Figure 7.4

of the quantity A, assuming that the file elements are in random order. The average

value of A, for a file of n elements, turns out to be $\sum_{i=2}^{n} 1/i$. For $n = 1000$, this sum is

6.5, so that for such a file, step 7 will have to be executed only six or seven times on average — a surprisingly small number. The mathematical analysis required to obtain this result is quite involved and is beyond the scope of this book. Note that to get a useful estimate of the time involved, as we have done, is comparatively straightforward

In the remaining two chapters we will be looking at some algorithms for problems of a numeric and non-numeric nature, and will return to the problem of algorithmic analysis there.

Problems

7.1 Give some examples of an algorithm, a procedure, and a heuristic.

7.2 Design an algorithm to find the prime factors of a given integer.

7.3 Consider the problem of analysing your algorithm. (A full analysis is likely to be very difficult, as it involves quite deep knowledge of number theory, but bounds on the run time should be fairly easy to obtain.)

7.4 It was stated in the text that an algorithm can often be built up from simpler algorithms. As an example, it is clear that an algorithm to sort a file into numerical order can be constructed from the algorithm already given which searched for the maximum element of a given file. One finds the maximum element, puts it at the head of the file, then sorts the remaining file for its maximum and so on. Construct such an algorithm.

7.5 Analyse the algorithm constructed in the previous question in order to determine its run time and storage requirements.

8. Sorting and Searching Algorithms

8.1 Introduction

A frequent problem in many scientific and business applications is the sorting of a file into some order, or the searching for a particular element in a file. The file may consist of a list of names to be *sorted* into alphabetical order, or a list of numbers to be sorted into numerical order. In *searching* we are seeking one or more distinct elements of a file. The processes of *sorting* and *searching* are clearly intimately connected, since it is possible to sort by searching for the first file element, then searching the remaining file for the next element, and so on. Also, it is in general easier to search a file that has been sorted than to search an unsorted file. There are vast number of sorting and searching algorithms, so the ideas of algorithmic analysis are needed in order that we can compare the different algorithms.

To decide on the *best* search or sort algorithm for a particular task usually involves compromise between various conflicting requirements. The two most commonly conflicting requirements are *storage space* and *execution time.* Also, some methods which are very efficient with a short file can become quite inefficient with a long file, and *vice versa.* Further, the method of sorting or searching clearly depends on the particular problem at hand. If one wishes to sort a file into a given order, all of whose elements are stored in high-speed core, then a certain class of algorithm should be used. This is called an *internal sorting* algorithm. If the file elements are stored on backing storage, such as a tape, drum or disc, then a different type of algorithm should be used, since only a subset of the file would be in high-speed core at any one time. This process is called *external sorting.*

In discussing sorting algorithms, we will consider the sorting of *n records* $r_1, r_2, \ldots,$ into some particular order. The order is specified by associating with each record a numerical *key*, so that the record r_i is associated with the key k_i. Usually two equal keys imply two identical records, though this is not invariably the case. The file is then sorted by arranging the keys in numerical sequence. In the following, when we discuss the

146

sorting of keys, it will be implicit that the associated records are also sorted. If the records are numerical, the keys may be the records themselves; that is, $k_i = r_i$. If the records are alphabetic, for example, a list of names, the keys can be derived from the records by associating a two digit integer with each letter; for example, $01 \equiv A$, $02 \equiv B, \ldots, 26 \equiv Z$. Or better, associating a three digit integer with each pair of alphabetic characters, so that $01 \equiv AA$, $02 \equiv AB$, $03 \equiv AC, \ldots, 676 \equiv ZZ$. The advantage of this second scheme, of course, is that only 3 digits are needed to key two alphabetic characters, while in the first scheme 4 digits are needed. Applying this transformation to the first N letters of the record — and right filling with zeros if any record is less than N characters long — results in a list of keys which, if sorted into ascending numerical order, gives a file whose records are sorted into alphabetical order. For example, given the following 9 names as records, the first two letters give the following keys under the two schemes already discussed:

Record	Key scheme A	Key scheme B
KEATS	1105	265
MAHONY	1301	313
WALLIS	2301	573
BRISLEY	0218	044
FICKER	0609	139
SMRZ	1913	481
GUTTMANN	0721	177
LAU	1201	287
GILES	0709	165

In either case, sorting these keys into ascending numerical order results in the associated records being sorted into alphabetical order.

Another common situation is that of the records being permutations of the integers — for example, a plant food company may be testing the effectiveness of some new products by applying them to a test garden bed in various orders, where the bed is prepared in different ways also. To particularize:

    Let 1 represent the planting of seed.
    Let 2 represent the watering of the soil.
    Let 3 represent the addition of SUPA-GRO plant food.
    Let 4 represent the addition of SMELLI-MUK manure.
    Let 5 represent the hoeing of the ground.
    Let 6 represent the addition of WUNDA-NITRATE to the soil.

There are six operations here that can take place in $6! = 720$ different sequences. Each sequence is uniquely represented by a permutation of the integers 1,2,3,4,5,6. Thus (5,1,6,2,4,3) represents the operation 5 (hoeing of the ground), followed by operation 1 (planting of the seed), followed by operation 6 (addition of WUNDA-NITRATE), and

so on. The results of the company's experiments, for example, the height, density, and
fruit-bearing capacity of the resultant plants, may be put into a file, along with the
permutation. Thus each record might consist of the experimental results and the identifying permutation. To obtain a suitable key for these records we require an algorithm
which associates a unique integer n with each permutation, such that $1 \leqslant n \leqslant 720$. That
is, given a set of distinct elements $(E_1, E_2, \ldots, E_m)$ we wish to find an integer
$n = f(E_1, E_2, \ldots, E_m)$ such that

$$1 \leqslant n = f(E_1, E_2, \ldots, E_m) \leqslant m!$$

with the condition that $f(E_1, E_2, \ldots, E_m) = f(F_1, F_2, \ldots, F_m)$ if and only if the two
sets $(E_1, E_2, \ldots, E_m)$ and $(F_1, F_2, \ldots, F_m)$ are identical.

An algorithm to find such an integer is given by Knuth (1969), and a slightly improved version is shown in the form of a flowchart (Figure 8.1).

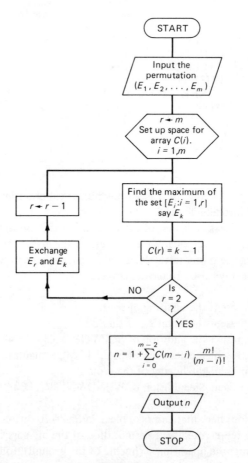

Figure 8.1

The effectiveness and uniqueness of this algorithm is proved by Knuth (1969).

To see how this algorithm works, it is instructive to take a simple example, such as the 3! = 6 permutations of the integers 1,2,3. For the permutation (2,1,3), we have $n = 3$,

E_1	E_2	E_3	r	k	$C(3)$	$C(2)$
2	1	3	3	3	2	–
2	1	3	2	1	2	0

Thus $n = 1 + C(3) + 3C(2) = 3$.

Similarly $(1,2,3) \Rightarrow 6$, $(3,1,2) \Rightarrow 1$, $(3,2,1) \Rightarrow 4$, $(1,3,2) \Rightarrow 5$, and $(2,3,1) \Rightarrow 2$.

In the following, we will assume that all file entries consist of a record with an associated numerical key. The file will be said to be *sorted* when the records are arranged in such a way that the keys are in ascending (or descending) numerical order. We will therefore restrict ourselves to considering methods of internal sorting of numerical keys.

3.2 Searching Algorithms

Linear Search

The most obvious search method is the *linear search*. In this method, the required key is compared with each key in turn until equality is achieved. For example, consider a file of 16 elements with the following keys: 09,72,63,19,02,41,18,94,04,23,37,67,59, 27,82,38, stored in array KEY(16). To search for key 32 say, which is not in the file, all 16 keys must be compared with the desired key. To search for a key that *is* in the file, a lesser number of comparisons will usually be necessary. In particular, if we assume the file keys are distinct and random, then a search for a particular key will obviously take on average $n/2$ comparisons for a file of length n. On the other hand, if we want say, the maximum or minimum of a file, which is found by comparing the first two keys for the required maximum (or minimum), then comparing the maximum (or minimum) of the first two keys with the third key, and so on, then $n-1$ comparisons will be necessary.

The above remarks apply to a linear search of an unsorted file. If the file is sorted, then a linear search for a specific key will take on average $n/2$ comparisons *whether the key is in the file or not*. The reason for this is, of course, that once we find that the current key is greater than the key for which we are searching, then a *match* will never be achieved and the search can be abandoned. To search for a maximum or minimum key of a sorted file does not take any comparisons at all, since we know that the maximum and minimum keys are at either end of the file.

A summary of the number of comparisons required by a linear search routine is shown below, for a file of length n elements.

Number of Comparisons in a Linear Search

	Unsorted file (keys assumed distinct and randomly distributed)	Sorted file (keys assumed distinct)
Search for a given key in the file	1 (minimum) $n/2$ (on average) n (worst case)	1 (minimum) $n/2$ (on average) n (maximum)
Search for a given key not in the file	n (exactly)	1 (minimum) $n/2$ (on average) n (maximum)
Search for largest or smallest key	$n-1$ (exactly)	zero (exactly)

Binary Search

A binary search is a method for searching a sorted file. It cannot be applied to an unsorted file, but is usually much faster than a linear search for sorted files.

To search for a given key, the file is divided into two equal portions (if possible) — if there is an odd number of keys in the file, the two portions will differ by 1 in the number of keys. It is then determined to which part the key belongs by comparing the given key with the last key of the first portion. This part is then again divided into two equal or approximately equal portions, and so on. For example, consider the file of 16 keys discussed previously, which is to be searched for a key 43. The binary search takes place as shown in the following diagram:

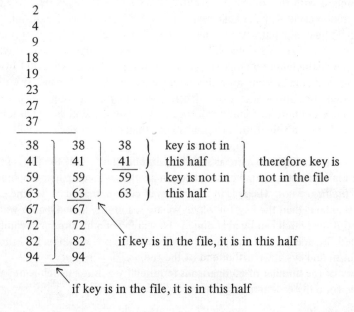

At each step, the size of the file is halved. If there are n keys in the file, then there exists a number j such that $2^j \geqslant n > 2^{j-1}$. The size of the file is reduced to 1 in j steps, that is $\lceil \log_2 n \rceil$, which is the least integer greater than or equal to $\log_2 n$. Therefore, to search for a key, whether or not the key is in the file, will take at most $1 + \lceil \log_2 n \rceil$ comparisons — where the 1 arises from the fact that a final comparison may have to be made at the end.

A FORTRAN program segment to search for key M in a sorted file IFILE of length $N > 1$ is given below:

```
          .
          .
          .
      IBEGN = 1
      IEND = N
    5 IMID = (IBEGN+IEND)/2
      IF (M−IFILE(IMID)) 10,20,30
   10 IF (IEND−IBEGN) 40,40,6
    6 IEND = IMID−1
      GO TO 5
   30 IF (IEND−IBEGN) 40,40,7
    7 IBEGN = IMID+1
      GO TO 5
   40 . . . we arrive here if the search fails
          .
          .
          .
   20 . . . we arrive here if the search succeeds, with
           M = IFILE(IMID)
```

Note that the above program can undoubtedly be made faster by, for example, replacing some arithmetic IF statements with logical IF statements, which are usually faster. Many other algorithms for searching a file exist, and a number of variations of these two methods are often used. For a fuller discussion see Knuth (1972).

To conclude our discussion of searching, we will consider a rather different type of algorithm, based on *hashing*.

Hashing

Another type of searching algorithm that is widely used in the construction of compilers and assemblers is based on a principle called *hashing*. Alternative names for this technique include scatter storage, random storage, key-transformation storage, and computed entry storage.

To demonstrate the principal idea, consider the following problem faced by a FORTRAN compiler. By the language rules, the maximum allowable number of identifiers is $26 \times (36^6 - 1)/35 \approx 1.62 \times 10^9$. Any given FORTRAN program will usually use only a small subset of the allowed set of identifiers. On accepting the program, the compiler sets up a so called *symbol table* that keeps track of all the identifiers used by the program. The table will contain a record for each identifier and each record will consist of the identifier, the type of the identifier, the address set aside for its storage, and possibly other information. As the FORTRAN program is scanned, new entries wil be inserted into the symbol table. Also, the compiler may wish to 'look up' a record in the table in order to determine certain information — such as the address of the identifier. Note that the table is built up dynamically, and its size is not known in advance.

One method of constructing such a table would be to set aside $1.62 \times 10^9 \times k$ word of core storage, where each record is assumed to take k words of core. In this way, space is set aside for all possible identifiers. This is of course a highly unsatisfactory solution due to its excessive space requirements. The usual solution to this problem is to provide a symbol table by *hashing*.

The basic idea in hashing is to provide a *hashing function* h, which maps all variable names onto the integers 1 through m. Many different identifier names may be hashed to the same integer. That is, the function h will not possess an inverse. It can be assume that it takes an approximately constant time to calculate $h(x)$, irrespective of x. There are numerous types of hashing function in the literature (see G. D. Knott (1975) for a review). One of the simplest is to square the binary representation of the identifier, extract the middle n digits and add 1. Thus $1 \leqslant h(x) \leqslant 2^n = m$.

The records are stored in an array H of size m, each entry of which contains a point to the head of a (possibly empty) list containing all variable names in the program that hash to the same value. Given p variable names, the expected length of each list will therefore be p/m. One usually arranges $m > p$, so the expected list length is <1. The resulting symbol table is shown below:

Given a variable name x, calculate $h(x) = k$ say, $1 < k \leqslant m$.
Insert x at the end of the list pointed to by $H(k)$.

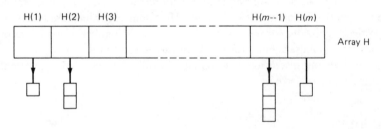

Figure 8.2

There are m lists, with list pointed to by $H(k)$ containing
all records which hash to k.

Thus the scheme effectively yields m lists instead of one list. Given a variable name one wishes to see if it is in the list. This is done by calculating $h(t)$ and searching the lis pointed to by $H(h(t))$. If the variable name is not in the list we insert the appropriate record at the end of the list pointed to by $H(h(t))$. Otherwise we skip the list, since the appropriate record is already there.

To search for an element takes one evaluation of the hashing function plus a search of a list of average length $p/m < 1$. In the worst case – if all identifiers hash to the same value – we have to search a file of length p, so we are no better off then if we had just used one list for all the variables, but on average we are better off by a factor of m ignoring the time taken to evaluate the hashing function.

There are a number of technical problems associated with hashing, including the selection of a suitable hashing function, minimizing the occurrence of two variables hashing to the same value – called *collision*, and the fact that we do not know the number of identifiers, p, in advance, which makes the choice of m difficult. One must therefore often be prepared to construct a sequence of hash tables of successively greater size. For a review of hash table methods, see Maurer and Lewis (1975).

3.3   Sorting Algorithms

We will now consider the problem of sorting a file into a given order, which is assumed to be in increasing or decreasing numerical order.

Sequential Sorting

A sequential sort is carried out by repeated application of a linear search. First the largest key is found and placed at the top of the file. This requires $n-1$ comparisons. The remaining $n-1$ keys are then searched for the next largest key, which requires $n-2$ comparisons. This key is then placed second on the file, and so on. The total number of comparisons is clearly

$$(n-1)+(n-2)+ \ldots +3+2+1 = n(n-1)/2 \approx n^2/2,$$

while the storage required is simply that occupied by the n records.

Insertion Sorting

A faster method of sorting, that takes no more storage space than a sequential sort, is an insertion sort. In this method we move keys from a given file to a sorted file, re-ordering at each step. This is best illustrated by an example. The first few steps in an insertion sort are shown in the table overleaf.

At the kth step, the kth key in the original file is inserted into its correct position in the sorted file of $k-1$ keys. This step is executed with $k = 1,2,3, \ldots ,n$ successively. The number of comparisons at each step depends on the search algorithm used to find the correct position for insertion. If a linear search is used, the average number of comparisons at the kth step is $(k-1)/2$. Thus the total average number of comparisons is

$$\sum_{k=1}^{n} (k-1)/2 = n(n-1)/4,$$

which is half the number of comparisons required by a sequential search.

Original file keys	New file (Step 1)	New file (Step 2)	New file (Step 3)	New file (Step 4)	New file (Step 5)	New file (Step 6)	New (Ste
09	9	72	72	72	72	72	7
72		9	63	63	63	63	6
63			9	19	19	41	4
19				9	9	19	1
02					2	9	1
41						2	
18							
94							
04							
23							
37							
67							
59							
27							
82							
38							

This algorithm can be made still faster, however, if we use a binary search to determine the correct position for the insertion of the kth key. In that case, the maximum number of comparisons is $1+\lceil\log_2(k-1)\rceil$, so that the total maximum number of comparisons is

$$\sum_{k=2}^{n}\{1+\lceil\log_2(k-1)\rceil\}\leqslant 1+\sum_{p=2}^{q}p2^{p-2}=1+(q-1)2^{q-1}$$

where $q=\lceil\log_2(n-1)\rceil+1$, so that the maximum number of comparisons is $\leqslant(q-1)2^{q}$ (The proof of this result is set as an exercise.) To compare these methods, let us take t case $n=1000$. A sequential sort takes approximately $10^6/2=500000$ comparisons. A insertion sort using a linear search for insertion takes, on average, half this number of steps; that is, 250000. An insertion sort using a binary search for insertion takes a max mum of about 10^4 comparisons, which is clearly an enormous saving.

Note, however, that an insertion sort does require more manipulation of the file elements than a sequential sort, due to the necessary insertion operations. However, the operation of insertion is normally faster than a comparison operation, and can often be neglected to a first approximation if an efficient data structure for the file ha been chosen. The required storage is again that occupied by the n file keys, since the sorted file can over-write the original file.

Radix Sorting

This method, also called *distribution sorting,* is an adaptation of the method used to sort punched cards into hoppers. In this method, each key is first sorted into a pocket according to its least significant digit. The keys are then collected in order and sorted according to the second digit, then collected, and so on. This is demonstrated by a radix sort of our earlier file keys:

Original file keys	1st sort		1st collection	2nd sort		2nd collection
09			41			02
72			72			04
63			02			09
19			82			18
02		0	63	09,04,02	00	19
41	41	1	23	19,18	10	23
18	82,2,72	2	94	27,23	20	27
94	23,63	3	04	38,37	30	37
04	4,94	4	37	41	40	38
23		5	67	59	50	41
37		6	27	67,63	60	59
67	27,67,37	7	18	72	70	63
59	38,18	8	38	82	80	67
27	59,19,9	9	09	94	90	72
82			19			82
38			59			94

When the *sort* is *collected,* by taking the keys from right to left in each 'pocket', the order of the keys in each digital position must be preserved. The number of sorts and collections depends on the maximum number of digits in a key, let us say k, which will usually be equal to $\lfloor \log_{10} n \rfloor$. There are n steps for each sort, so the total number of comparisons' is nk, where we have neglected the collection process, the time for which s small compared to the time taken by the sorting process. Note that a 'comparison' for this sorting method means determining the appropriate pocket. It is thus likely to be more time consuming than an ordinary comparison, in which we determine if x is greater or less than y.

Radix sorting is a particularly useful method when $k \ll n$. $11n$ storage locations are required by this method, since n locations are required to store the original file, and these same n locations can be used at each collection stage. At the sorting stage, each of the 10 digital positions must be able to hold up to n keys. This method can be re-fined by efficient data structuring so that a lesser number of storage locations are required, but in general it is a method that is fairly 'core hungry'.

Bubble Sort

Another fairly inefficient sorting method is a bubble sort. In this method, the first two keys are compared, and interchanged if necessary. The second and third keys are compared and then interchanged if necessary, and so on until the $(n-1)$th and nth keys are compared and interchanged if necessary. This is repeated with the first $(n-1)$ keys, then the first $(n-2)$ keys, and so on. The method is demonstrated below for a short file:

72	19	19	19	19	19	09
19	72	63	63	63	09	19
63	63	72	09	09	63	63
09	09	09	72	72	72	72

first pass   *second pass*   *third and final pass*

As can be seen from the above example, the smallest key, 09, 'bubbles up' through the file to the top, hence the name *bubble sort.* Note that the keys are sorted at any pass when no interchange takes place. From the description of the sort procedure, it can be seen that the number of comparisons required to sort n items is at most

$$\sum_{j=1}^{n-1} j = n(n-1)/2.$$

Since this is a maximum, a bubble sort may be faster than a sequential sort, though the same number of storage locations (n) are required.

Merge Sort

Merge sorting is quite an efficient way of sorting and a number of variations of this basic method exist. We will consider a *two-way merge sort.* In this method, pairs of keys are examined in turn and ordered with the smaller key placed first. After one pass, the initial file of n keys consists of $n/2$ files of length 2 keys. Pairs of files of length 2 are then combined to produce $n/4$ files of length 4, and so on, until only one file remains. This is demonstrated in Figure 8.3 opposite.

For a file of length n, the number of passes is clearly $\lceil \log_2 n \rceil$.

Note that in this method, in addition to the operations of comparing and interchanging keys, it is also necessary to *merge* two sorted files to produce a single longer sorted file. To merge two *ordered* files, each of length p say, requires at most $2p$ comparisons. The merging takes place as follows. Assume we wish to merge two ordered arrays A(I) and B(I) of length M and N elements respectively to produce a new array C(I) of length $M+N$. First compare A(1) and B(1), placing the smaller in C(1). Then if the smaller element were A(1), compare A(2) to B(1), placing the smaller element in C(2). If the smaller element were B(1), then compare B(2) to A(1), placing the smaller element in C(2). Continue this process until all elements of either A or B have been placed in C. If all elements of A have been placed in C, then place the remaining elements of B in C, and similarly, if all elements of B have been placed in C, then place the

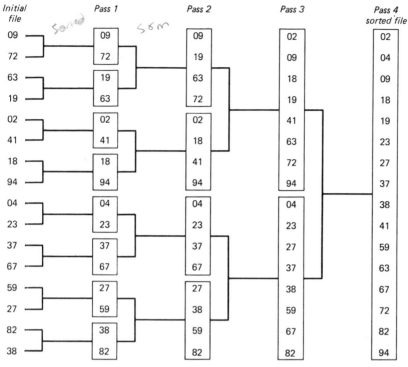

Figure 8.3

remaining elements of A in C. A FORTRAN subroutine to carry out this merge operation
is given below:

```
SUBROUTINE MERGE(A,B,C,M,N)
DIMENSION A(1),B(1),C(1)
I = 1
J = 1
K = 0
1 K = K+1
IF (A(I).GT.B(J)) GO TO 2
C(K) = A(I)
I = I+1
IF (I.LE.M) GO TO 1
GO TO 4
2 C(K) = B(J)
J = J+1
IF (J.LE.N) GO TO 1
DO 3 L = I,M
K = K+1
C(K) = A(L)                    (cont.)
```

```
3  CONTINUE
   RETURN
4  DO 5 L = J,N
   K = K+1
   C(K) = B(L)
5  CONTINUE
   RETURN
   END
```

In each pass of a merge sort, there are at most n comparisons to be made, and, as we have seen, there are $\lceil \log_2 n \rceil$ passes. The total number of comparisons is thus approximately $n \lceil \log_2 n \rceil$. With $n = 1000$, the number of comparisons is therefore ≈ 10000, a similar result to that obtained by an interchange sort with binary search for insertion.

The two-way merge sort just considered can clearly be generalized to a p-way merge. The first step is to divide the file into n/p ordered subfiles of length p, then to merge these in groups of p to give n/p^2 ordered subfiles of length p^2. This process continues until only one ordered file remains. A difficulty with $p > 2$ is that the merge operation becomes more complicated than for $p = 2$. It can be shown that a p-way merge requires $(p-1)n \lceil \log_p n \rceil$ comparisons, and $2n$ storage locations.

Merge-exchange Sorting (Shell Sorting)

One particularly efficient way of sorting, both with regard to storage requirements and number of comparisons, is the method of merge-exchange sorting or shell sorting. Knuth (1972) calls this method *diminishing-increment sorting*. It consists of a mixture of merging and exchanging operations. On the first pass, the n items are grouped into $n/2$ subfiles of length 2, comparing and ordering a_1 and $a_{\lfloor n/2 \rfloor + 1}$ (note that $\lfloor x \rfloor$ is the largest integer not greater than x), then a_2 and $a_{\lfloor n/2 \rfloor + 2}$ and so on, so that after this step, $a_1 \leqslant a_{\lfloor n/2 \rfloor + 1}, a_2 \leqslant a_{\lfloor n/2 \rfloor + 2}, \ldots$. The file is then divided into $n/4$ subfiles, each of length 4, with the first subfile containing the elements $\{a_1, a_{\lfloor n/4 \rfloor + 1}, a_{\lfloor n/2 \rfloor + 1}, a_{\lfloor 3n/4 \rfloor + 1}\}$ etc. These subfiles are then ordered, then merged into groups of 8 and so on. This is best illustrated by an example (see Figure 8.4 (opposite)).

In this example we have used the sequence 8,4,2,1. Other sequences could just as well be used, and indeed, a great deal of work has been done to determine the best sequence. A good choice in general is the sequence $1,4,13,40,121,364, \ldots, (3^s - 1)/2$. For the sequence $1,2,4,8, \ldots, 2^s$ the number of comparisons is around $1.6n \lceil \log_2 n \rceil$, while only n storage locations are required. The mathematical analysis of this particular algorithm is quite intricate. The work on this problem up to 1972 is discussed by Knuth (1973).

Heapsort and Quicksort

Two other fast methods of sorting are called *heapsort* and *quicksort*. In *heapsort*, the file must first be 'heapified' — that is, put in the form of a heap. A heap is a binary tree-like structure, with each son being less than or equal to its father. Once the file is partially ordered into a heap, it can be sorted very quickly. It can be shown that about

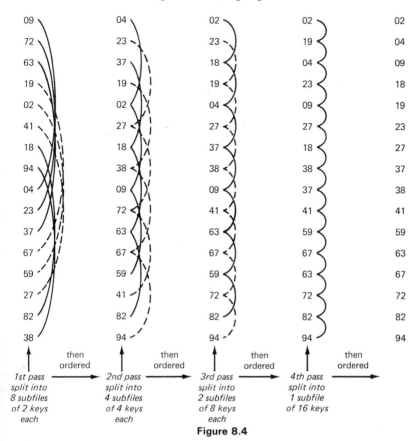

09	04	02	02	02
72	23	23	19	04
63	37	18	04	09
19	19	19	23	18
02	02	04	09	19
41	27	27	27	23
18	18	37	18	27
94	38	38	38	37
04	09	09	37	38
23	72	41	41	41
37	63	63	59	59
67	67	67	67	63
59	59	59	63	67
27	41	72	72	72
82	82	82	82	82
38	94	94	94	94
1st pass split into 8 subfiles of 2 keys each	then ordered 2nd pass split into 4 subfiles of 4 keys each	then ordered 3rd pass split into 2 subfiles of 8 keys each	then ordered 4th pass split into 1 subfile of 16 keys each	then ordered

Figure 8.4

$\lceil \log_2 n \rceil$ comparisons are required to heapify the file, and $n \lceil \log_2 n \rceil$ comparisons are required to sort the heap.

Quicksort is a recursively defined sorting algorithm that takes $1.4n \lceil \log_2 n \rceil$ comparisons on average but has a worst case behaviour of n^2 comparisons.

Both these algorithms are fairly complex, though very efficient, and the reader is referred to Knuth (1973) for details of both these and other sorting algorithms not discussed here.

In the table on page 160 we summarize the storage and time requirements of the sorting algorithms we have discussed.

If the maximum number of digits in the file key k is sufficiently small, the *radix sort* the fastest method, with only nk comparisons. On the other hand, its storage requirements are fairly heavy, with $11n$ storage locations being needed. Also, bear in mind that 'comparison' for the radix sort is a more time consuming operation than for the other methods.

If storage is at a premium, the *shell sort, heapsort* or *quicksort* are amongst the best methods. A FORTRAN subroutine to carry out a shell sort is given by Berztiss (1971).

bubble sort or *sequential sort* has the advantage of programming ease and minimum storage, though both become unacceptably slow with increasing file length.

Method	Number of comparisons $\leqslant N.$ N	Storage requirements
Sequential sort	$n(n-1)/2$	n
Insertion sort:		
(a) with linear search	$n(n-1)/4$	n
(b) with binary search	$1+(q-1)2^{q-1}$ with $q = \lceil \log_2(n-1) \rceil + 1$	n
Radix sort	nk (k = no. of digits in file key)	$11n$
Bubble sort	$n(n-1)/2$	n
p-way merge sort	$(p-1)n\lceil \log_p n \rceil$	$2n$
Heapsort	$2n\lceil \log_2 n \rceil$	n
Quicksort	$1.4n\lceil \log_2 n \rceil, n^2$ (on average) (at worst)	$n+\lceil \log_2 n \rceil$
Merge-exchange (shell) sort	$1.6n\lceil \log_2 n \rceil$ (empirically)	n

There are many other algorithms for internal sorting, but most of them are variatic of the principal types described here. A fuller discussion is given in Knuth (1973) and the other references cited in the bibliography for this chapter, on page 208.

Problems

8.1   Translate the flowchart on page 148 into a FORTRAN or ALGOL program. Te it by systematically generating all $n!$ permutations of $\{1,2,3,4, \ldots ,n\}$ for $n = 3,4,5$, and 6, and see that they are all uniquely numbered by the algorithm.

8.2   The FORTRAN program segment for a binary search may not be entirely satis-factory. Study it carefully, see if it works in every case, and if not, correct it.

8.3   Verify that $\displaystyle\sum_{k=2}^{n} \{1+\lceil \log_2(k-1) \rceil\} \leqslant 1+\sum_{p=2}^{q} p2^{p-2} = 1+(q-1)2^{q-1}$,

where $q = \lceil \log_2(n-1) \rceil +1$.

8.4   Write a program to do an insertion sort with binary search.

8.5   One characteristic of the sorting algorithms we have discussed has not been men tioned. If a file does not have all the keys distinct, are those keys which are the same left in the sorted file in the order in which they existed in the original file If so, we call the algorithm *stable*. Clearly, sequential sorting is stable. Heapsort is unstable. Are the other sorting algorithms we have discussed stable or unstab

9. Random Numbers and Simulation

9.1 Introduction

In many computer applications it is necessary to generate a sequence of random numbers. The principal application is in *simulation,* where a computer program is written to simulate some event or process. We will consider an example of this later in this chapter, when we study the operation of a railway level crossing. One piece of input information that must be simulated is the arrival of a stream of cars at random times. Thus, a source of random numbers will be required for this simulation program.

Random numbers are also useful in *game theory,* a trivial example being the throwing of a die, or a game of 'two up'. In these applications random numbers in the range 1—6 and 1—4 respectively would be used to simulate all possible occurrences. That is the toss of a six-sided die reveals a face between 1 and 6 spots. In 'two up', the four equally likely outcomes of tossing two coins (head-head, head-tail, tail-head and tail-tail) are represented by the digits 1 to 4.

In numerical analysis, a source of random numbers may be used to estimate, for example, the value of an integral which cannot be determined analytically. Consider the integral

$$I = \int_a^b f(x) \, dx,$$

where f is a known function, sketched in Figure 9.1, with $0 < f(x) < 1$ for $x \in [a,b]$.

If we generate two sequences of uniformly distributed random numbers $\{x\}$ and $\{y\}$ satisfying $x \in [a,b]$ and $y \in [0,1]$, then take successive pairs of elements (x,y) as coordinates in the $x - y$ plane, and count (1) the total number of such pairs, N say, and (2) the number of points that lie in the region bounded by $x = a$, $x = b$, $y = 0$, and $= f(x)$, N_I say, then the value of the integral I is approximated by $(N_I/N)(b - a)$. The accuracy of the approximation increases roughly as the square root of N. Such methods are known as *Monte Carlo* methods.

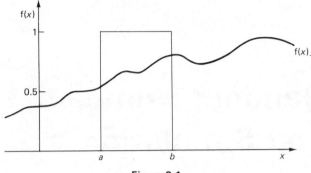

Figure 9.1

These are just some of the more obvious applications of random numbers, and in this chapter we will consider a number of algorithms for generating random numbers. Before we do this, let us review some results in statistics which will be used later.

9.2   Review of Some Statistical Concepts

Distributions

The key concept that will concern us is that of a *distribution.* A distribution can be characterized either by its *density function* or its *cumulative distribution function.* The last is often called simply a *distribution function.*

Consider first a discrete distribution, such as the distribution of a population by age. The density function $p(x)$ can take values only for discrete values of age, x_i, and $p(x_i)$ represents the probability that $x = x_i$. Obviously, we must have $\sum_i p(x_i) = 1$ and $p(x_i) \geqslant$ for all i.

An example of a discrete distribution function is the *Poisson distribution,* defined

$$p(n) = e^{-\lambda} \frac{\lambda^n}{n!} \quad n = 0,1,2, \ldots ; \lambda > 0.$$

Then $p(0) = e^{-\lambda}$, $p(1) = \lambda e^{-\lambda}$, and $\displaystyle\sum_{n=0}^{\infty} p(n) = \sum_{n=0}^{\infty} e^{-\lambda} \frac{\lambda^n}{n!} = e^{-\lambda} \cdot e^{\lambda} = 1.$

For example, if a radioactive sample causes a Geiger-counter to register λ times per second on average, then the probability of n registrations during any second is

$$e^{-\lambda} \frac{(\lambda)^n}{n!}.$$

A continuous density function, also denoted by $p(x)$, is an obvious generalization of a discrete density function. The probability that x takes on a value between x_i and x_j given by

$$\int_{x_i}^{x_j} p(x) \, dx.$$

We must have $\int_{-\infty}^{\infty} p(x)\, dx = 1$ and $p(x) \geqslant 0$.

The simplest continuous distribution is the *uniform distribution*, defined by

$$p(x) = \frac{1}{b-a} \qquad a \leqslant x \leqslant b$$

$$= 0 \qquad\qquad \text{otherwise.}$$

Clearly the two conditions for a continuous distribution are satisfied, since

$$\text{(a) } p(x) \geqslant 0 \quad \text{and} \quad \text{(b) } \int_{-\infty}^{\infty} p(x)\, dx = \int_{a}^{b} p(x)\, dx = 1.$$

Another continuous distribution of great significance is the *normal distribution*, which is defined by the density function

$$p(x) = \frac{1}{\sigma \sqrt{2\pi}} e^{-[(x-\mu)/\sigma]^2/2},$$

with $\sigma > 0, \mu > 0$. Clearly $p(x) \geqslant 0$, and to evaluate $\int_{-\infty}^{\infty} p(x)\, dx$ we first make the substitution $t = (x-\mu)/\sigma$. Then

$$I = \int_{-\infty}^{\infty} p(x)\, dx = \frac{1}{\sigma\sqrt{2\pi}} \int_{-\infty}^{\infty} e^{-t^2/2} \sigma\, dt.$$

Thus

$$I^2 = \frac{1}{2\pi} \int_{-\infty}^{\infty} e^{-t^2/2}\, dt \cdot \int_{-\infty}^{\infty} e^{-s^2/2}\, ds$$

$$= \frac{1}{2\pi} \int_{-\infty}^{\infty} \int_{-\infty}^{\infty} e^{-(s^2+t^2)/2}\, ds\, dt$$

$$= \frac{1}{2\pi} \int_{0}^{\infty} \int_{0}^{2\pi} r e^{-r^2/2}\, d\theta\, dr = 1,$$

where we have transformed from cartesian to polar co-ordinates in the last line. As we shall see, the constants σ and μ turn out to be the standard deviation and mean, respectively, of the distribution. The normal distribution is that which many natural events are assumed to follow in statistical trials. It is the best known and most frequently used continuous probability distribution.

Another important continuous distribution is the exponential distribution, which is defined by the density function

$$p(x) = \lambda e^{-\lambda x} \quad x \geqslant 0, \lambda > 0$$
$$= 0 \qquad x < 0.$$

Again, $p(x) \geqslant 0$ by inspection, and $\int_{-\infty}^{\infty} p(x)\,dx = \int_{0}^{\infty} \lambda e^{-\lambda x}\,dx = \lambda/\lambda = 1$. The exponential distribution is appropriate to a number of naturally occurring queuing situations; for example, the time interval between cars at an intersection, or the time interval between arrival of orders at a factory.

There are, of course, numerous other distributions, but these are the most common, and also those that we will be using in the remainder of this chapter.

Another way of characterizing a distribution, the *distribution function,* is defined by

$$F(x) = \sum_{x_i \leqslant x} p(x_i)$$

for a discrete distribution, and

$$F(x) = \int_{-\infty}^{x} p(t)\,dt$$

for a continuous distribution. $F(x)$ is then the probability that $(x \leqslant X)$ for the random quantity X. In the discrete case, the graph of the distribution function will look like a series of horizontal line segments, while in the continuous case the distribution function will be a smooth, non-decreasing curve. Clearly, $F(-\infty) = 0$ and $F(\infty) = 1$. For the four distributions we have considered, the distribution functions are:

(i) Poisson distribution function

$$F(X) = \sum_{X_i = 0}^{\lfloor X \rfloor} e^{-\lambda} \frac{\lambda^{X_i}}{X_i!} \qquad \text{e.g. } F(2.7) = e^{-\lambda} + \lambda e^{-\lambda} + \tfrac{1}{2}\lambda^2 e^{-\lambda}$$

where $\lfloor X \rfloor$ is the largest integer not greater than X. Then $F(2.7)$ is the probability that the value of the random variable distributed according to the given density function does not exceed 2.7.

(ii) Uniform distribution function

$$F(x) = \int_{-\infty}^{x} \frac{1}{b-a}\,dt = \begin{cases} 0 & \text{if } x \leqslant a \\ x-a & \text{if } a < x < b \\ 1 & \text{if } x \geqslant b \end{cases}$$

The graph of this function is as shown in Figure 9.2.

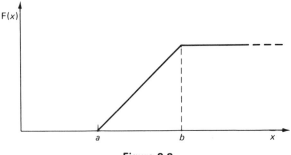

Figure 9.2

(iii) Normal distribution function

This is given by

$$F(x) = \frac{1}{\sigma\sqrt{2\pi}} \int_{-\infty}^{x} e^{-(t-\mu)^2/2\sigma^2} \, dt$$

which obviously can be obtained from tables of the error function,

$$\text{erf}(z) = \frac{2}{\sqrt{\pi}} \int_{0}^{z} e^{-t^2} \, dt \text{ by using } F(x) = \tfrac{1}{2}\left\{1 + \text{erf}\left(\frac{x-\mu}{\sqrt{2\sigma}}\right)\right\}$$

(iv) Exponential distribution function

This is given by

$$F(x) = \int_{-\infty}^{x} p(t) \, dt = \int_{0}^{x} \lambda e^{-\lambda t} \, dt = 1 - e^{-\lambda x}.$$

Expected Values

The *expected value* of any function of x, $f(x)$ say, where x is distributed according to a given density function, is written $E\{f(x)\}$ and is defined by

$$E\{f(x)\} = \begin{cases} \displaystyle\sum_{\text{all } x_i} p(x_i)f(x_i) & \text{for the discrete case} \\[2em] \displaystyle\int_{-\infty}^{\infty} p(t)f(t) \, dt & \text{for the continuous case} \end{cases}$$

The *mean* of a distribution is denoted by $\mu = E\{x\}$ and the *variance* $\sigma^2 = E\{(x - \mu)^2\} =$

$E\{x^2\} - \mu^2$. The mean and variance of the four distributions we have discussed are given in the table below:

Distribution	Mean	Variance
Poisson	λ	λ
Uniform	$(b + a)/2$	$\dfrac{(b - a)}{12}$
Normal	μ	σ^2
Exponential	$1/\lambda$	$1/\lambda^2$

9.3 Conversion from One Distribution to Another

Most of the random number generators we will discuss give a uniformly distributed sequence of numbers between 0 and 1. For many applications we require random numbers which are distributed normally, exponentially or according to some other distribution law. We therefore want to use a source of uniformly distributed data to generate data distributed according to some other distribution law.

In principle, this is straight-forward, since any continuous distribution function

$$y = F(x) = \text{probability that the random quantity } X \leqslant x$$

has an inverse function G, such that

$$x = F^{-1}(y) = G(y).$$

Thus, given a uniformly distributed quantity U, between 0 and 1, the quantity $G(U)$ is distributed according to the distribution law $F(x)$.

The problem is now to find this inverse function G. For an exponential distribution, this is easy since

$$F(x) = 1 - e^{-\lambda x}$$

so that

$$G(x) = F^{-1}(x) = -\frac{1}{\lambda} \ln[1 - x].$$

Given a uniformly distributed random variable, x, such that $0 \leqslant x \leqslant 1$, an exponentially distributed random variable, w, with mean $1/\lambda$, can be constructed by the formula

$$w = -\frac{1}{\lambda} \ln[1 - x].$$

As a slight simplification, we observe that if x is uniformly distributed between 0 and 1, so is $1 - x$, so we can generate an exponentially distributed random variable from $w = -\dfrac{1}{\lambda}\ln(x)$. A FORTRAN subprogram to achieve this conversion is given below:

```
     FUNCTION EXPRND(X,XLAM)
C    GIVEN A RANDOM VARIABLE X BETWEEN 0 AND 1, WHICH IS
C    UNIFORMLY DISTRIBUTED, EXPRND IS AN EXPONENTIALLY
C    DISTRIBUTED RANDOM VARIABLE WITH MEAN=1/XLAM AND
C    VARIANCE=1/XLAM**2
     IF (X.GT.0.0) GO TO 100
C    IF X=0, SET EXPRND TO LARGEST FLOATING POINT NUMBER
C    THAT THE MACHINE CAN STORE
     EXPRND = 1.0E+76
     RETURN
100  EXPRND = -ALOG(X)/XLAM
     RETURN
     END
```

To achieve a normally distributed random number is a considerably more difficult problem, since we have to invert — at least in principle — the distribution function

$$F(x) = \frac{1}{\sigma\sqrt{2\pi}} \int_{-\infty}^{x} e^{-(t-\mu)^2/2\sigma^2} \, dt.$$

A simple, closed form expression for the inverse F^{-1} clearly does not exist. One approach due to Hastings (1955) is to use a polynomial approximation to the inverse function, and the algorithm which achieves this is as follows. Given a uniformly distributed random variable X lying between 0 and 1, an *approximately* normally distributed random variable Y, with mean 0 and variance 1, can be generated as follows:

> Initialize $c_0 \leftarrow 2.515517; c_1 \leftarrow 0.802853; c_2 \leftarrow 0.010328;$
> $\qquad d_1 \leftarrow 1.432788; d_2 \leftarrow 0.189269; d_3 \leftarrow 0.001308$
> IF $0 \leqslant X \leqslant 0.5$ THEN $t \leftarrow (-2\ln X)^{\frac{1}{2}}$ ELSE $t \leftarrow (-2\ln(1-X))^{\frac{1}{2}}$
> $\qquad Z \leftarrow t - (c_0 + c_1 t + c_2 t^2)(1 + d_1 t + d_2 t^2 + d_3 t^3)$
> IF $0 \leqslant X \leqslant 0.5$ THEN $Y \leftarrow -Z$ ELSE $Y \leftarrow Z$
> Output Y

A FORTRAN subprogram to achieve this transformation is given below:

```
     FUNCTION XNORM(X)
     DATA C0,C1,C2,D1,D2,D3/2.515517,0.802853,0.010328,1.432788,
    10.189269,0.001308/
     IF (X*(1.0-X)) 10,20,30
10   WRITE (2,100)X
100  FORMAT (22H X IS OUT OF RANGE, X=,E12.4)
     STOP
20   IX = X
     XNORM = (-1**(IX+1))*1.E76              (cont.)
```

```
C     THIS SETS XNORM TO LARGEST MACHINE NUMBER IF X=1 AND TO
C     LARGEST MACHINE NUMBER IF X=0
      RETURN
   30 IF (X.GT.0.5) GO TO 40
      T = SQRT(-2.0*ALOG(X))
      Z = T-((C2*T+C1)*T+C0)/(((D3*T+D2)*T+D1)*T+1.0)
      XNORM = -Z
      RETURN
   40 T = SQRT(-2.0*ALOG(1.0-X))
      Z = T-((C2*T+C1)*T+C0)/(((D3*T+D2)*T+D1)*T+1.0)
      XNORM = Z
      RETURN
      END
```

Another method for transforming from a uniform to a normal distribution is the polar method, due to Box, Muller, and Marsaglia. This method requires two *independent* uniformly-distributed random variables X_1 and X_2 (both lying in the range [0,1]), and produces two independent normally-distributed random variables Y_1 and Y_2, with zero mean and variance 1, as follows:

$$\text{REPEAT BEGIN} \qquad\qquad Z_1 \leftarrow 2X_1 - 1;$$
$$Z_2 \leftarrow 2X_2 - 1;$$
$$A \leftarrow Z_1{}^2 + Z_2{}^2$$
$$\text{END}$$
$$\text{UNTIL } A < 1;$$
$$T \leftarrow (-2\ln(A)/A)^{\frac{1}{2}}; Y_1 \leftarrow Z_1 t; Y_2 \leftarrow Z_2 t$$
$$\text{Output } Y_1, Y_2$$

The proof that this method works is given in Knuth (1969).

A very clever and fast algorithm to transform from a uniformly distributed random variable to a normally distributed random variable has been written by Marsaglia and MacLaren. It requires auxiliary storage of about 100 entries, and is best programmed in assembly language, since it requires a lot of bit manipulation. For further details the reader is referred to the original paper, or the very readable account in Knuth (1969).

As our final transformation, to generate a set of Poisson distributed variables with mean λ, given a source of random variates $\{X_n\}$ uniformly distributed between 0 and 1, we may use the following algorithm:

$$N \leftarrow 0; q \leftarrow 1; p \leftarrow e^{-\lambda}$$
$$\text{REPEAT BEGIN input } X;$$
$$q \leftarrow q \cdot X;$$
$$N \leftarrow N+1;$$
$$\text{END}$$
$$\text{UNTIL } q < p$$
$$N \leftarrow N-1;$$
$$\text{Output } N$$

A FORTRAN subprogram to implement this algorithm is given below:

```
        FUNCTION NPOIS(XLAMDA)
        NPOIS = 0
        Q = 1.0
        P = EXP(-XLAMDA)
      5 CALL NURND(X)
C       THIS SUBROUTINE PROVIDES A UNIFORMLY DISTRIBUTED
C       RANDOM VARIATE X
        Q = Q*X
        IF (Q.LT.P) GO TO 10
        NPOIS = NPOIS+1
        GO TO 5
     10 RETURN
        END
```

This algorithm, and a terse proof of its validity, is given by Knuth (1969), who also discusses a faster algorithm, due to P. Kribs, which achieves the same purpose. However, Krib's algorithm is more difficult to program and requires an auxiliary storage table. Both algorithms work very well for small values of λ, since then N is usually very small. The higher the value of λ, the slower the run time for the algorithm.

Having completed our review of some of the statistical results we shall need, we now consider methods for generating random numbers.

9.4 Tests for Randomness

We shall now look at schemes for generating *uniformly distributed* random numbers in the interval [0,1].

The earliest computer oriented method for generating random numbers was proposed by von Neumann and Metropolis in 1946. Loosely known as the *middle square* method, the idea was to take the square of the previous random number and to take the middle digits of the square as the new random number. For example, if we are generating 6 digit random numbers and the first number is 631704, its square is the 12 digit number 399049943616, so the next random number is given by the middle 6 digits 049943, and so on. (Each number so determined must be divided by 999999 to give a number in the range [0,1].)

This method raises the question of the meaning of 'random number', since numbers generated in this way are not random, in the sense that each number is uniquely determined by its predecessor. Compare this to the rolling of a die, when each number is truly random, assuming a true die.

At this time, nearly all computer methods of generating random numbers suffer from the same defect as the middle square method; that is, that they are not truly random. For this reason, such numbers are often called *pseudo-random numbers*. For most practical purposes they can be considered to be purely random numbers, provided they satisfy a number of statistical tests for randomness. In the following, we shall not distinguish between random and pseudo-random numbers, but will just refer to random numbers.

One feature of any method of random number generation that depends only on the previous number is that of limited *periodicity*. The periodicity of a sequence of random numbers is the number of random numbers in a cycle before the numbers start to repeat If we are generating random numbers with m decimal digits, the periodicity is *at most* 10^m, since there are only 10^m different m digit decimal integers, and since each number depends only on its predecessor, once one number repeats itself, so does the whole cycle. Note that 10^m is the *maximum* periodicity of such a sequence of random numbers, and in practice it is often very much less. The major fault of the middle square method is that the period is often very short, since the length of the period depends critically on the starting number. For example, if we are using 4 digit numbers and take 3100 as our starting number, the cycle goes 3100,6100,2100,4100,8100,6100 . . . , so that we immediately degenerate into a cycle of period 4. Futher, once the number 0 is obtained as a random number in the middle square method, it obviously sticks there since its square is zero etc.

Let us look at some statistical tests that a sequence of random numbers must pass to be acceptable. Obviously we require the period to be as long as possible, though this is not really a statistical test. We will assume in the following that the numbers are uniformly distributed in $[0,1]$. This suggests one obvious test, that the mean of a sequence of random numbers should be 0.5 with standard deviation $1/\sqrt{12}$.

A more stringent test is the *frequency test*. In this test we are checking to see that the numbers are uniformly distributed, so that the probability of finding a number in any sub-interval of $[0,1]$ is the same as the probability of finding a number in any other sub interval of equal length. We first generate a set of N random numbers, $n_1, n_2, \ldots, n_N$. If we divide the unit interval into k equal sub-intervals, then the average number of random numbers in each sub-interval should be N/k. Let r_i $(i = 1, 2, \ldots, k)$ denote the actual number of random numbers in the sub-interval $[(i-1)/k, i/k]$. Then form the statistic

$$\chi^2 = \frac{k}{N} \sum_{i=1}^{k} \left(\frac{N}{k} - r_i \right)^2 .$$

This is the 'chi-squared' statistic of the observed quantities r_i. By comparing this number with a table of chi-square values we obtain a probabilistic answer to our question. For example, if we took 1000 random numbers and divided the unit interval into 20 equal divisions, we would look up a table of χ^2 statistics with $20 - 1 = 19$ degrees of freedom Such a table is given in *Handbook of Mathematical Functions* (edited by M. Abramowi and I. A. Stegun), table 26.8, for example. We find there that the 10 per cent and 90 per cent probability points occur with values of χ^2 of 27.20 and 11.65 respectively. This means that χ^2 can be expected to be greater than 11.65 for 90 per cent of experiments but greater than 27.20 for only 10 per cent of the experiments. We can consider our results satisfactory then if χ^2 lies between these two figures about 80 per cent of the time. The test should therefore be repeated for a number, M say, of sequences of N random numbers. If we repeat the test 100 times say, we would expect about 80 values of χ^2 to lie within the two limits quoted above.

There is another — and better — frequency test which uses the Kolmogorov-Smirnov test. This is a technique for finding a confidence band for the distribution function of a continuous variable. Details of the test are given in any book on statistics, and its application to a frequency test on methods of random number generation is discussed by Knuth (1969).

Another important test is the *serial test,* which tests the degree of randomness between successive pairs of numbers. In this test we are looking for correlations — or more accurately for the absence of correlations — between successive pairs of numbers. Proceeding as for the frequency test, the unit interval is divided into k sub-intervals. Then let r_{ij} denote the number of random numbers n_l $(l = 1,2, \ldots ,N - 1)$ which satisfy $(i - 1)/k \leqslant n_l < i/k$ and $(j - 1)/k \leqslant n_{l+1} < j/k$ for each value of i and j in the interval $[1,k]$. There are $N - 1$ successive pairs of numbers n_l, n_{l+1}, and if the numbers are truly random, the probability that $n_l \in [(i - 1)/k, i/k)$ is $1/k$ with the same probability that $n_{l+1} \in [(j - 1)/k, j/k)$. Thus r_{ij} can be expected to have, on average, the value $(N - 1)/k^2$. Given the quantities r_{ij} we compute the χ^2 statistic

$$\chi_2{}^2 = \frac{k^2}{N-1} \sum_{i=1}^{k} \sum_{j=1}^{k} \left(\frac{N-1}{k^2} - r_{ij} \right)$$

for the entire sample of N random numbers. This can then be compared to a table of χ^2 statistics with $k^2 - 1$ degrees of freedom. As with the frequency test, this should be repeated for a number of samples of random numbers.

There are a number of other tests used to test a sequence of random numbers, which include the following: The *lagged product test* is used to test the absence of correlation between numbers separated by a *lag* of k in the sequence. The statistic

$$\sum_{i=1}^{N-k} n_i n_{i+k}$$

is formed for a range of values of k. The statistic can be shown to be approximately normally distributed for uniformly-distributed random numbers, a result which can be tested by a chi-squared test.

The *runs test* is used to check for *runs up* and *runs down* in the sequence of random numbers. For example, if we are studying random digits in the range 0–9, and we obtain the following sequence

$$1 \; 7 \; 3 \; 9 \; 4 \; 6 \; 2 \; 7 \; 1 \; 0 \; 4 \; 6 \; 9 \; 3 \; 8 \; 5$$

then we see there are sequences of length 2,2,2,2,1,4,2,1, in a *runs up* test and sequences of length 1,2,2,2,3,1,1,2,2 in a *runs down* test. The number of runs of different lengths may then be compared with expected values in a fairly complicated way.

The *poker test* is a test for five or more digits in a random number. The digits in each group of 5 are related to poker hands, and the occurrence frequency of pairs, threes of a kind, full houses, and others, are compared with expected values, usually by a chi-squared test.

The *maximum test* considers the maximum element of a set of N random numbers n in the range $[0,1]$. The maximum element n_{max} is determined for a large number M of sets of N random numbers. The quantity n_{max}^N should then be uniformly distributed over $[0,1]$. This can be tested by applying the Kolmogorov-Smirnov test to n_{max}, with the distribution function $F(x) = x^N$.

There are still other statistical tests that can be applied to a random number generator but any method of random number generation that passes all the foregoing tests is likely to be satisfactory. Further details of these and other tests are given by Knuth (1969).

9.5   Linear Congruential Method

As we have seen, the middle square method of generating random numbers is unsatisfactory. The most widely used method today is the linear congruential method, due to D. Lehmer. In this method, each random number n_k is determined from its predecessor n_{k-1} according to the formula

$$n_k = (An_{k-1} + B) \bmod C, \qquad k \geqslant 1,$$

where $A, B,$ and C are given constants. The success of the method depends on the choice of constants $A, B,$ and C. For example, if $A = B = 5$, $C = 10$, and $n_0 = 3$, then the following sequence of random numbers would be generated: $3, 0, 5, 0, 5, 0, \ldots$ which would not satisfy many statistical tests for randomness!

The maximum possible period of the sequence of random numbers is clearly C. It is thus desirable to choose C to be as large as possible. It is customary to choose C to be 2^m (on a binary machine) or 10^m (on a decimal machine) where m is the number of bits or digits in a word. Thus on an ICL 1900 series with a 24 bit word, one bit of which is reserved for the sign, $m = 24$, so that $C = 2^{24}$, while with an IBM 360 series machine we would choose $m = 32$. The advantage of this choice is that the modulus operation takes place automatically, since on multiplication the right-most m bits of the product are stored as the result. To be specific, if, with $m = 24$, the value of $An_{l-1} + B$ is $k < 2^{24}$, then $k \bmod 2^{24} = k$, so the modulus operation is performed correctly. If $k \geqslant 2^{24}$, then k can be written $k = k_1 2^{24} + k_2$ where k_1 and k_2 are positive integers and $0 \leqslant k_2 < 2^{24}$. The number k_2 would be stored as the result, and is obviously equal to $k(\bmod 2^{24})$.

To ensure a period of length C, it is also necessary to choose A and B judiciously. The choice of A should be made such that if C is a power of 10, then $A(\bmod 200)$ should be chosen equal to 21. If a binary machine is being used, so that C is a power of 2, then $A(\bmod 8)$ should be chosen equal to 5. Further, A should be chosen to lie between $C/100$ and $C - \sqrt{C}$. The digits of A should also be chosen to have some haphazard pattern, such as 71394265, and not, say, 25252525. To be really sure that our choice of A is suitable, the resultant sequence of random numbers should be subjected to the *spectral test,* which is a fairly intricate algorithm due to Coveyou and MacPherson (1965). We will not discuss the test here, but for anyone who is serious about writing a good random number generator, the spectral test is described both in the original paper and in Knuth (1969), and should be used.

The constant B should be an odd number (if C is a power of 2) or not divisible by 5 (if C is a power of 10). Further, B/C, should be chosen to have the approximate value $\frac{1}{2}-\sqrt{3}/6 \approx 0.2113248654$. This choice minimizes the *serial correlation*; that is, the correlation between two successive random numbers.

The starting value n_0 may be chosen quite randomly; for example, the current date and time. Alternatively, the last random number used on the previous run could be used at the beginning of the next run.

Finally, note that this method generates a random integer n_i in the range $n_i \in [0, C-1]$. To obtain a random, uniformly distributed number in the range $[0,1)$ it is necessary to divide the random integer by $C-1$. If the desired range is $[0,1]$, the divisor should be C; if the desired range is $(0,1)$, then $n_i + 1$ should be divided by $C + 1$, and if the desired range is $(0,1]$, then $n_i + 1$ should be divided by C.

9.6   The MacLaren–Marsaglia Method

The linear congruential method with constants $A, B,$ and C as chosen above should be satisfactory for most applications. For those applications requiring a 'super random number generator', the following algorithm due to MacLaren and Marsaglia could be used. The method requires as input two independent sets of random numbers obtained by using the linear congruential method twice, with different multipliers A and constants B. An auxiliary storage area of about 100 locations is also needed. The algorithm then proceeds as follows. Let $\{n_i\}$ and $\{m_i\}$ be two independent sequences of random numbers. Let there be an auxiliary storage area containing 100 elements, $\{l_i: i = 1,100\}$. Fill the table $\{l_i\}$ with the first 100 elements of $\{n_i\}$; that is, $l_i = n_i$ for $i = 1,2, \ldots, 100$. Set $i = 0$ to initialize.

```
Initialize i ← 0
REPEAT BEGIN i ← i+1;
             N ← n_i;
             M ← m_i;
             j ← ⌊100M/C⌋ ; (where C is the modulus used in deriving the
                            sequence {M_i} using the linear congruential
                            method)
             output l_j;
             l_j ← N
      END
UNTIL you have enough random numbers
```

The output sequence l_j will in general be a considerably more random sequence than either of the input sequences $\{n_i\}$ or $\{m_i\}$.

In the next section we will see how random number generators are used in a typical, simulation problem.

9.7 Adamstown Railway Level Crossing: a Simulation Problem

As an example of a typical simulation problem, we will study the railway level crossing at Glebe Road, Newcastle, New South Wales. This is sketched in Figure 9.3 and as can be seen it consists of a two-lane two-way carriage-way, crossing two railway lines:

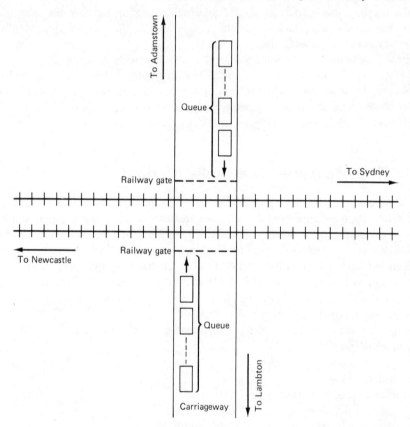

Figure 9.3

This crossing is manually controlled. Each time a train wishes to cross the road, the traffic must be stopped by closing the railway gates. Eventually the level crossing will be replaced by an overpass for vehicular traffic, but until that distant day, the authorities wish to know what delays and queue lengths are likely to arise at the crossing, so that the trains may be scheduled in such a way as to minimize the delays.

As a result of traffic counter measurements, the hourly flow of traffic along Glebe Road is known, and to simplify the problem, for this example, we will assume that each day has the same hourly distribution. This is, of course, an over-simplification, since on weekends the traffic flow pattern will be considerably different to that which occurs during the week. We also assume that the current traffic flow measurements can be extrapolated for the next few years.

The trains are assumed to arrive at the crossing at fixed times, as specified by some timetable. A slight deviation from the scheduled times must also be allowed for, since the trains may possibly be running late, or even a few minutes early (unlikely as this may sound, it is possible).

The gates are closed to traffic a fixed time T_1 seconds before the arrival of the train, and remain closed if another train is due to arrive within a fixed time interval T_2 seconds of the first train. The train takes T_T seconds to cross the road. The vehicles are assumed to form a single queue on each side of the crossing. We will study just one queue, that going to Adamstown say, as the queue travelling in the other direction can be studied in exactly the same way.

Let $f_1, f_2, f_3, \ldots, f_{24}$ be the mean number of vehicles per hour for each of the 24 hours of the day. The time interval between vehicles can be assumed to be exponentially distributed, with mean time interval between vehicles of $1/f_i$ for the ith hour of the day.

The trains are scheduled to arrive at fixed times $t_1, t_2, \ldots, t_n$ during the day, but may deviate from these times by an amount δt, where δt can be assumed to be normally distributed with mean 0 and variance s^2 say. Each vehicle can be assumed to take a time T_V seconds to cross the railway tracks.

The basic strategy of our simulation is shown on the first flowchart (Figure 9.4). Firstly, all relevant times and parameters are initialized. We will take our unit of time as the second, so that the day has elapsed when $T = 86,400$. The entire train timetable for the day is also read in at this stage. The four key events are: (1) the arrival of a vehicle at time TA, (2) the departure of a vehicle at time TD, (3) the closing of the gates at time TC, and (4) the opening of the gates at time TO. As the day evolves, we determine which event occurs next and then take appropriate action, re-setting some or all of the four key times as appropriate.

The type of information we seek as output to the problem depends on the parameter(s) we wish to optimize. In this situation we may wish to minimize the maximum queue length, or the average queue length. We shall not worry too much about this point, but we will calculate, say, the total number of vehicles, the total delay time of all vehicles, the ratio of these quantities, which is, of course, the average delay time, and the queue length at any moment. We shall also print out a message informing us of the current event; i.e., the gate opening or closing, or a vehicle arriving or departing. For simplicity we shall assume that at the start of the day, that is, at midnight plus 1 second, the gate is open and there is no vehicle waiting to cross, so we set TO and TD to infinity - which ensures that the next event is not the opening of the gate (since it is already open), or the departure of a vehicle (since no vehicle has yet arrived).

In order to calculate the arrival schedule of trains and cars, we need a source of normally distributed and exponentially distributed random numbers. Presumably, we could provide this by using the linear congruential method with appropriate constants, which provides a source of uniformly distributed random numbers. This sequence of uniform distribution could then be transformed into an exponential and a normal distribution by one of the methods already discussed. We shall assume that an appropriately distributed source of random numbers is available. The overall flowchart is then as shown on page 176.

The first event is either the arrival of a vehicle or the closure of the gates, prior to the arrival of a train. If the first event is the arrival of a vehicle, the flowchart is followed

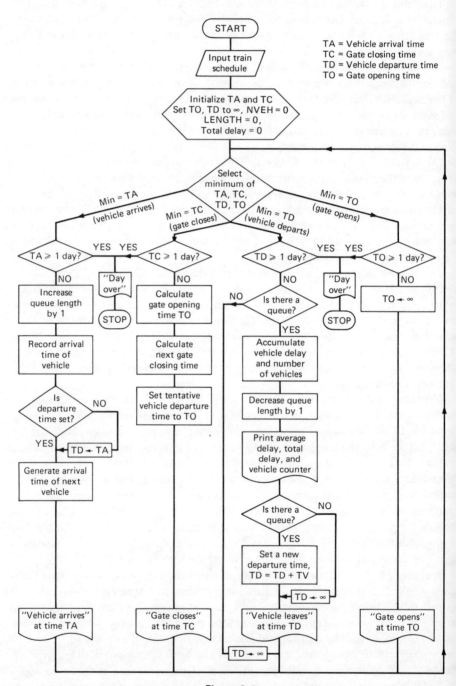

Figure 9.4

long the left-hand branch. Having established that we are still in the first day, the
vehicle queue length is increased by 1 and the arrival time of the vehicle is recorded. If a
departure time has not been set, it is now set. A departure time will not have been set
only if: (1) the queue length is now 1 and (2) the gate is open. As we shall see, under
any other circumstances a departure time will have been set. We therefore set the departure time TD to the arrival time TA. The next vehicle arrival time is set (involving a call
to the random number generator), the message 'vehicle arrives' is printed out, and the
program returns to the major decision box. At this stage TA has been re-set and TD has
been set, either in this branch of the flowchart or at an earlier stage.

If the first event is the closing of the gates, flow proceeds along the second branch of
the flowchart. The gate opening time is calculated, which depends of course on the train
schedule. One complication is that the gates may have to stay shut for the passing of
more than one train. The next gate closing time is also calculated, as is the arrival time
of the following train, since this updates the arrival time used in the calculation of the
gate opening time in the next call to this section. The next vehicle departure time is
tentatively set to the gate opening time TO. If there is no vehicle waiting to cross at this
time, it will be ignored. The message 'gate closes' is printed out and the program returns
to the major decision box.

If the next event is the departure of a vehicle and the day is not yet over, the appropriate action depends on whether there is a queue. If there is no queue, we just re-set TD
to infinity (or a machine approximation thereof), which ensures that the *next* event will
not be the departure of a vehicle. Since there is no queue, it is impossible for a vehicle
to depart. If there is a queue, however, we calculate the delay, which is accumulated,
and increase the total number of vehicles by 1. Then we decrease the queue length by 1
and print out the total delay, total vehicle throughput, and average delay at this stage.
Then we check if there is a queue *now*; that is, after one vehicle has passed. If so, we set
the new vehicle departure time. If there is no queue at this stage, we again set TD to
infinity, and in either case we print out the message 'vehicle leaves' and return to the
main decision box.

Finally, if the opening of the gate is the next event, the following gate opening time
is set to infinity, since the *next* gate opening time depends on the gate closing time, and
is set in the 'gate closing' branch.

Before formulating the problem with expanded and detailed flowcharts, we will
require the following variables:

TA = the time the next vehicle arrives
TD = the time the next vehicle departs
TC = the next time the gates will close
TO = the next time the gates will open
t_{DUE} = the time the next train is due
$t_{\text{DUE}+1}$ = the time the next train but one is due
$r_e(f_i)$ = a random exponential variate with mean = $1/f_i$
$r_n(s)$ = a random normal variate with mean 0 and variance = s^2
$NVEH$ = the number of vehicles
$NDEL$ = delay time of vehicle

$TQ(I)$ = arrival time of Ith vehicle in queue – we must assign an array of appropr
size to hold the largest queue

$LENGTH$ = current queue length

$TR(IT)$ = scheduled arrival time of ITth train – this array must be large enough to
hold a 24 hour train timetable

$TDEL$ = total delay experienced by all vehicles

T_1 = time interval before arrival of train that gates close

T_T = time taken for train to cross the road

T_2 = if next train is due within T_2 seconds of present train, the gate stays
closed until both trains pass

T_V = average time taken for a vehicle to cross the road

G: a dichotomic variable. When $G = 0$ the gate is open, when $G = 1$ the gate
is closed.

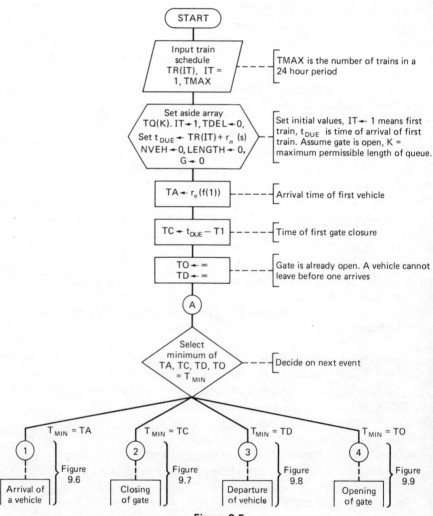

Figure 9.5

The expanded flowcharts are shown in Figure 9.5 to 9.9, and should be compared with the earlier flowchart (Figure 9.4) to see how the finer details are attended to. To change this flowchart into a programming language should be fairly straightforward, and is left as an exercise. This algorithm was developed by stepwise refinement, but we have only shown an intermediate and the final levels of the design.

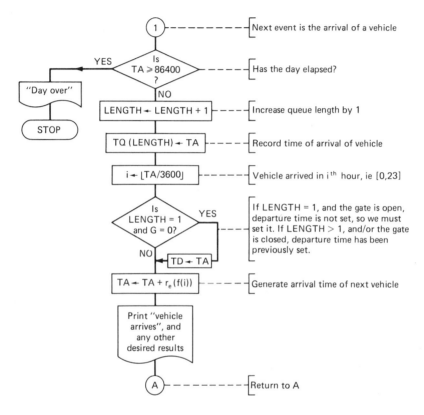

Figure 9.6

Each event that takes place during the day requires a substantial amount of programming, so that with a flow of thousands of cars and perhaps 100 trains, it is clear that the simulation program will take a long time to run. This highlights the major problem associated with computer simulation, that while quite complex and realistic simulations can be formulated algorithmically without a great deal of effort, the successful execution of the algorithm often requires a prohibitive amount of time.

This completes our discussion of distribution theory, random number generation, and simulation. All these topics are explored in greater depth in the various references cited on page 209.

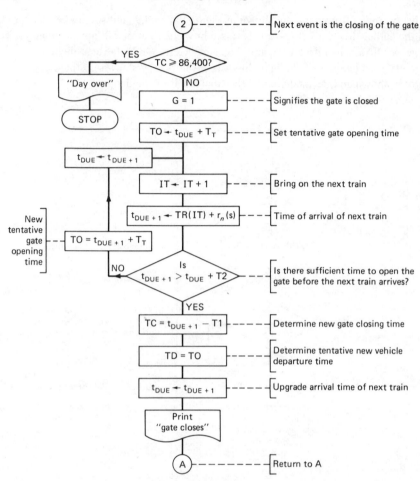

Figure 9.7

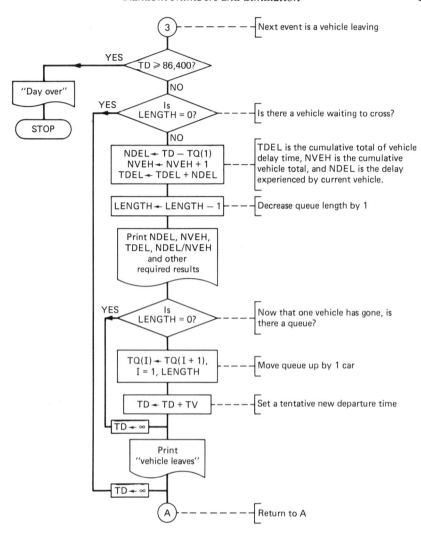

Figure 9.8

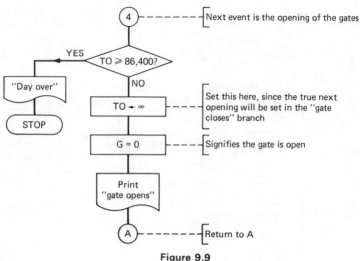

Figure 9.9

Problems

9.1 Calculate the mean and variance of the four distributions considered. That is, verify the results given in the table on page 166.

9.2 Write a FORTRAN or ALGOL subroutine to implement the polar method for generating normally distributed random variables.

9.3 Generate a sequence of uniformly distributed random numbers in the range [0,1] on your local computer using the linear congruential method.

9.4 (*For enthusiasts only.*) Select reasonable traffic volumes, consult a train timetable. For a crossing at a local railway station, simulate the railway level crossing in an appropriate programming language, then send a copy of your results to the appropriate local newspaper, and a letter of abuse or praise to the Railways Board.

Solutions to Selected Problems

Chapter 1

1.1 If $X > 1$, take the integer M such that $0 < X - M \leqslant 1$. Calculate e^{X-M} from the power series expansion, and multiply by e^M, which is readily evaluated by multiplying $\underbrace{e \times e \times e \ldots \ldots}_{M \text{ factors}}$

M factors

The case $X < -1$ is treated similarly. The test for six significant figure accuracy is now slightly more difficult to apply, as the factor e^M must be incorporated into the test. For $-1 \leqslant X \leqslant 1$ the power series may be summed directly.

1.2

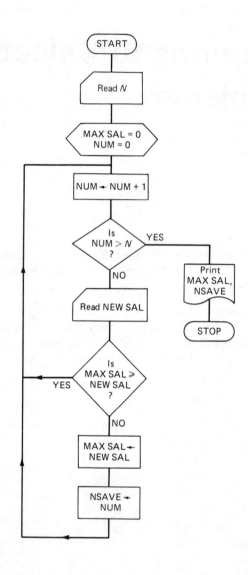

1.3

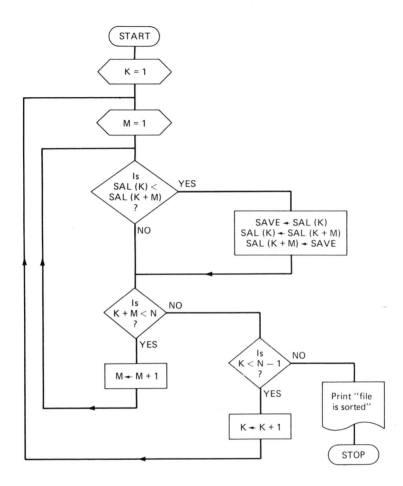

1.4

IQ $<$ 90	Y	N	N	N	N	N
90 $\leqslant$ IQ $<$ 115	N	Y	Y	N	N	N
115 $\leqslant$ IQ $<$ 165	N	N	N	Y	Y	N
IQ $\geqslant$ 165	N	N	N	N	N	Y
Average mark $\geqslant$ 65	–	–	Y	Y	N	–
Average mark $\geqslant$ 80	–	N	Y	–	N	–
Class 3 institution	X					
Class 2 institution		X			X	
Class 1 institution			X	X		
Egghead College						X

1.5

1. HAS N BEEN READ?	N	Y	Y	Y	Y
2. IS NUM $>$ N?		Y	N	N	N
3. IS MAXSAL $\geqslant$ NEWSAL?				Y	N
READ N	X				
SET MAXSAL=NUM=0	X				
READ NEWSAL			X		
MAXSAL=NEWSAL					X
NSAVE=NUM					X
ADD 1 TO NUM	X			X	X
GO TO 2	X			X	X
GO TO 3			X		
PRINT MAXSAL,NSAVE		X			
STOP		X			

1.6

Is C true?	Y	N
Do S	X	
Skip S		X

IF C DO S

1.

Is C true?	Y	N
Do S	X	
Go to 1	X	
Skip S		X

WHILE C DO S

1.

Has i been set? Is i.sign(C) ≤ B.sign(C)?	N –	Y Y	Y N
Set i ← A	X		
Do S		X	
Increment i ← i + C		X	
Go to 1	X	X	
Skip S			X

FOR i=A TO B IN STEPS OF C DO S

1.

Has S been done? Is C true?	N –	Y N	Y Y
Do S	X	X	
Skip S			X
Go to 1	X	X	

REPEAT S UNTIL C

1.7

Read n
Initialize max to zero
FOR I=1 TO N IN STEPS OF 1 DO
    BEGIN
        Read SALARY
        IF SALARY > MAX THEN BEGIN
                MAX ← SALARY;
                INDEX ← I
            END
    END
Print MAX, INDEX

1.8 This problem is solved by a sequence of IF-THEN-ELSE structures, in a manner similar to the solution of the airline scheduling problem given on page 16.

Chapter 2

2.1

```
NEW EUCLID
10 READ A,B
15 IF A = 0 THEN 97
20 IF B = 0 THEN 97
25 LET C = ABS(A)
30 LET D = ABS(B)
35 LET Q = INT(C/D)
40 LET R = C−D*Q
45 IF R = 0 THEN 65
50 LET C = D
55 LET D = R
60 GO TO 35
65 PRINT "THE G.C.D. OF",A,"AND",B,"IS",D
70 GO TO 10
75 DATA 589,−171,5648,8657432
80 STOP
97 PRINT "STUPID DATA,A=",A,"B=",B
98 GO TO 10
99 END
```

2.2

```
NEW PRIMES
10 DIM P(100)
12 PRINT "THE 1ST PRIME IS 2"
15 LET P(1) = 2
17 PRINT "THE 2ND PRIME IS 3"
20 LET P(2) = 3
25 LET I = 2
30 LET K = 5
35 LET L = 2
40 REM CALCULATE FRACTIONAL PART OF K/P(L)
45 LET F = K/P(L)−INT(K/P(L))
50 IF F<1.0E−5 THEN 90
55 IF P(L)>SQR(K) THEN 70
60 LET L = L+1
65 GO TO 45
70 P(I+1) = K
75 PRINT "THE",I+1,"TH PRIME IS",P(I+1)
80 IF I = 99 THEN 98
85 I = I+1
90 K = K+2
95 GO TO 35
98 STOP
99 END
RUN
```

The following output is then produced:

THE 1TH PRIME IS 2
THE 2TH PRIME IS 3
THE 3TH PRIME IS 5
THE 4TH PRIME IS 7
THE 5TH PRIME IS 11

.
.
.

THE 100TH PRIME IS 541

This program uses the facts that, other than 2, all primes are odd, and that to test a number N for primeness, one need only attempt to divide it by primes $\leqslant \sqrt{N}$. If there is a non-zero remainder in each case, the number is a prime.

2.3

```
NEW SIMPS
10 DEF FNA(X) = EXP(-X*X)
15 LET S = FNA(0)+FNA(1)+4*FNA(0.01)
20 FOR I = 2 TO 98 STEP 2
25 LET X = I*0.01
30 LET S = S+4*FNA(X+0.01)+2*FNA(X)
35 NEXT I
40 PRINT "ERF(1)=",S/150.0/SQR(3.141592654)
45 STOP
RUN
```

The output is then

ERF(1) = 0.842701

2.4

```
010  REMARK: PROGRAM TO SOLVE QUADRATIC EQUATIONS
020  READ A,B,C
025  PRINT "*********************************************************************"
030  PRINT "A =";A;"B =";B;"C =";C
040  IF A = 0 THEN 110
050  IF C = 0 THEN 120
060  IF ABS((B*B)/(4.0*A*C)) > 1.0E4 THEN 130
070  LET D = B*B - 4.0*A*C
080  IF D < 0 THEN 140
085  REMARK: THE ABOVE STATEMENTS, AND THE FOLLOWING "GO TO"
090  REMARK: STATEMENTS SET UP THE INDEX "I" FOR THE CASE (I)
095  REMARK: CONTROL STRUCTURE
100  I=5
105  GO TO 150
110  I=1
115  GO TO 150
120  I=2
125  GO TO 150
```

(cont.)

```
130 I=3
135 GO TO 150
140 I=4
150 ON I GO TO 160,200,200,250,300
160 IF B = 0 THEN 400
170 PRINT "EQUATION IS LINEAR"
180 PRINT "ROOT =", -C/B
190 GO TO 20
200 REMARK: B**2≥4*A*C, SO SOLUTION IS OBTAINED BY EXPANDING
205 REMARK: SQRT (D)
210 REMARK: WHERE D IS DEFINED IN STATEMENT 70. THIS SECTION
220 REMARK: ALSO HANDLES THE CASE WHEN C = 0.
230 PRINT "ROOT 1=", -C/B, "ROOT 2 =", -B/A + C/B
240 GO TO 20
250 REMARK: THE FOLLOWING IS THE CASE OF COMPLEX ROOTS
255 S=SQR(-D)
260 PRINT "COMPLEX ROOT CASE"
270 PRINT "REAL PART ROOT 1 =";-B/2/A;"IMAG. PT. ROOT 1 =";S/2/A
280 PRINT "REAL PART ROOT 2 =";-B/2/A;"IMAG. PT. ROOT 2 =";-S/2/A
290 GO TO 20
300 LET S = SQR(D)
310 PRINT "REAL ROOTS"
320 PRINT "ROOT 1 =", (S-B)/2/A," ROOT 2 =", (-S-B)/2/A
330 GO TO 20
340 DATA 8.0,-22.0,15.0
350 DATA 0.0,4.0,-5.0
360 DATA 1.0E-8,1.0E-4,1.0E-8
370 DATA 8.0,4.0,0.5
380 DATA 3.0,1.0,2.0
390 DATA 0.0,0.0,0.0
400 END
```

Chapter 3

3.1
```
      SAVE = VEC(K)
      DO 100 I = 2,K
      KK = K−I+1
      VEC(KK+1) = VEC(KK)
  100 CONTINUE
      VEC(1) = SAVE
```

3.2
```
      DIMENSION IMAT(5,5)
      DATA IMAT/25*0/
      DO 200 J = 2,4,2
      DO 200 I = 1,5,2
      IMAT(J,I) = I*J
      IMAT(I,J) = I+J
  200 CONTINUE
```

3.3 (a)

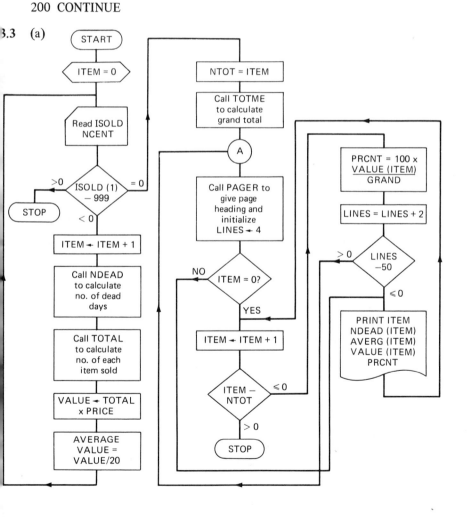

(b) ITEM needs to be initialized to zero.
1100 ITEM = 0

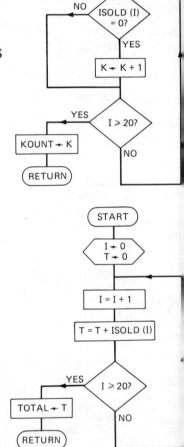

(c) C THIS PROGRAM PROVIDES A
C SALES SUMMARY FOR UP TO
C 200 ITEMS, FOR EACH ITEM AN
C IDENTIFYING NUMBER, THE
C NUMBER OF DAYS NO SALES
C WERE MADE, THE AVERAGE
C DAILY SALES, THE TOTAL
C MONTHLY VALUE AND THE
C PERCENTAGE THIS REPRESENTS
C OF TOTAL SALES VALUE IS
C PRINTED OUT. ON EACH PAGE
C THE GRAND SALES TOTAL IS
C GIVEN

(d) ```
     FUNCTION KOUNT(ISOLD)
     DIMENSION ISOLD(20)
     K = 0
     DO 100 I = 1,20
     IF (ISOLD(I).EQ.0)K = K+1
100 CONTINUE
     KOUNT = K
     RETURN
     END
```

(e)  ```
     FUNCTION TOTAL(ISOLD)
     DIMENSION ISOLD(1)
     T = 0.0
     DO 100 I = 1,20
     T = T+ISOLD(I)
100  CONTINUE
     TOTAL = T
     RETURN
     END
```

(f) 2800 AVERG(ITEM) = VALUE(ITEM)/20.0
3000 GO TO 2000

(g) 5300 IF(LINES−50) 5600,5600,5400
5600 WRITE(NPRINT,5700) ITEM, NDEAD(ITEM),
        1   AVERG(ITEM),VALUE(ITEM),PRCNT

(h)   After calling PAGER for the first time, ITEM is increased by 1, so that it
       has the value NTOT+1. The decision box following this terminates the
       program. This can be rectified by initializing ITEM to zero at point A   in
       the flowchart, which corresponds to inserting the statement ITEM = 0
       immediately after statement number 4200.

(i)   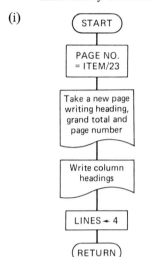 Requires ITEM, GRAND, and LINES, since
       this routine also initializes LINES.

(j)   TOTME calculates the total sales of *all* items for the month, so requires the
       value of sales for all items. TOTAL calculates how many of each item has
       been sold. This is independent of the value of any item. Hence the two
       function subroutines are different, and so are their arguments.

```
4       F(X) = EXP(-X*X)
  100   FORMAT (8H1ERF(1)= ,F16.9)
        S = F(0.0)+F(1.0)+4.0*F(0.01)
        DO 200 I = 2,98,2
        X = FLOAT(I)/100.0
        S = S+4.0*F(X+0.01)+2.0*F(X)
  200   CONTINUE
        ANS = S/150.0/SQRT(3.1415926536)
        WRITE (6,100)ANS
        STOP
        END
```

```
3.5        INTEGER A,B,C,D,Q,R
      1 READ (1,100)A,B
    100 FORMAT (2I10)
        IF (A.EQ.0) GO TO 99
     10 C = IABS(A)
        D = IABS(B)
     15 Q = C/D
        R = C-D*Q
        IF(R. EQ. 0) GO TO 30
     20 C = D
        D = R
        GO TO 15
     30 WRITE(6,200)A,B,D
    200 FORMAT (14HOTHE G.C.D. OF, I7, 4H AND,I7,3H IS,I7)
        GO TO 1
     99 STOP
        END

3.6        DIMENSION NP(100)
        DATA NP/2,3/
        K = 5
        L = 3
        DO 100 I = 3,100
      5 NPL = NP(L)
        NREM = K-(K/NPL)*NPL
        IF(NREM .EQ. 0) GO TO 90
     10 NROOT = SQRT(K)
        IF(NPL-NROOT .GT. 0) GO TO 40
     20 L = L+1
        GO TO 5
     90 K = K+2
        L = 3
        GO TO 5
     40 NP(I) = K
        K = K+2
        L = 3
    100 CONTINUE
        WRITE(6,200) (I,NP(I),I=1,100)
    200 FORMAT(4H THE,I3,12H TH PRIME IS,I4)
        STOP
        END
```

Chapter 4

4.1 (a) $(A + 4)/2/B$ (b) $A\uparrow(j+1)$

    (c) $(A + B)\times(C+D)/(E+F)$ (d) $\ln(y+\text{sqrt}(y\times y - 1))$

4.2 (a) $p:= (A/(A+1.5))\uparrow 3.5$ (b) $GC:= h/4\times(4\times R - h)/(3\times R - h)$

    (c) $l:= m:= (a\times a+1\cdot (c-d)\uparrow 2)/(b\uparrow 3+(c+a)\uparrow 2\times e)$

4.3 $a = 2, b = 0$

4.4 (a) **if** $Y \geqslant X$ **then begin if** $Z \geqslant Y$ **then** $MAX:= Z$ **else** $MAX:= Y$
               end else if $Z \geqslant X$ **then** $MAX:= Z$ **else** $MAX:= X$;

    (b) **if** $x \leqslant -2$ **then** $FUN:= -2$ **else if** $x \leqslant -1$ **then** $FUN:= -1+x/2$
        else if $x < 0$ **then** $FUN:= x\times(2.5+x)$ **else** $FUN:= 0$;

4.5
```
real sum,term; integer i;
sum:= 3↑(−4/3);
a:= i:= 2;
for i:= a×i while term ⩾ 10⁻⁸ do
begin term:= (i+1)↑(−4/3);
        sum:= sum+term
end;
```

4.6 (a)
```
T[1] := 1.5; fae:= 2;
for n:= 2 step 1 until p do
begin m:= 2×n;
        fae:= fae×(m−1)×m;
    T[n] := (m+1)/fae
end;
```

    (b)
```
for q:= −5 step 1 until 5 do
begin R:= 1+(2×q)↑2
    for S:= 0 step 1 until 5 do
    T[q,s] := R×(2×q×s−2)
end;
```

4.7
```
procedure triang(a,b,c,area); value a,b,c;
real a,b,c,area;
    begin real s,t; s:= (a+b+c)/2;
        t:= s×(s−a)×(s−b)×(s−c);
      area:= sqrt(t)
    end;
```

```
4.9    begin integer n,i,j,k,l;
        start: n := read;
               if n = 0 then go to end;
               begin integer array latin [1:n,1:n] ;
                  for i := 1 step 1 until n do
                  latin [1,i] := read;
                  for i := 1 step 1 until n do
                     begin l := latin [1,latin[1,i] ] ;
                     for j := 2 step 1 until n do
                     begin k := i+j−1−nxentier ((i+j−2)/n);
                     latin [j,latin[1,k] ] = l
                     end this completes the construction of the latin square,
                     and the next section prints it out
                     end; paperthrow;
               for i := 1 step 1 until n do
                  begin newline(2);
                  for j := 1 step 1 until n do
                  print (latin[i,j] ,5,0)
                  end;
               go to start;
               end;
        end:end;
```

Chapter 5

5.1

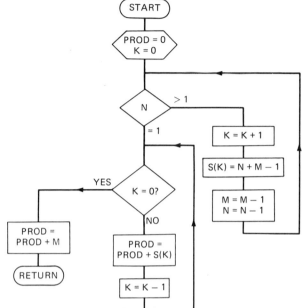

```
   INTEGER FUNCTION PROD(NA,MA)
   INTEGER S(500),PROD
   N=NA
   M=MA
   PROD=0
   K=0
   IF (N.LT.0) GO TO 70
   IF (N.GT.0) RETURN
10 IF (N.GT.1) GO TO 40
30 IF (K.EQ.0) GO TO 65
   PROD=PROD + S(K)
   K=K−1
   GO TO 30
40 K=K+1
   S(K)=N+M−1
   N=N−1
   M=M−1
   GO TO 10
65 PROD=PROD+M
   RETURN
70 WRITE(6,80) N
80 FORMAT('0 N MUST BE POSITIVE, N =',I8)
   RETURN
   END
```

5.2   In this example we have taken a slightly different approach. We have set up a 2-dimensional stack, storing the N and M values. At each step we remove the bottom pair of elements from the stack, apply the definition of the function to these two elements, and then shift the stack down to fill up the gap created. The flowchart is given below, with the corresponding FORTRAN program opposite.

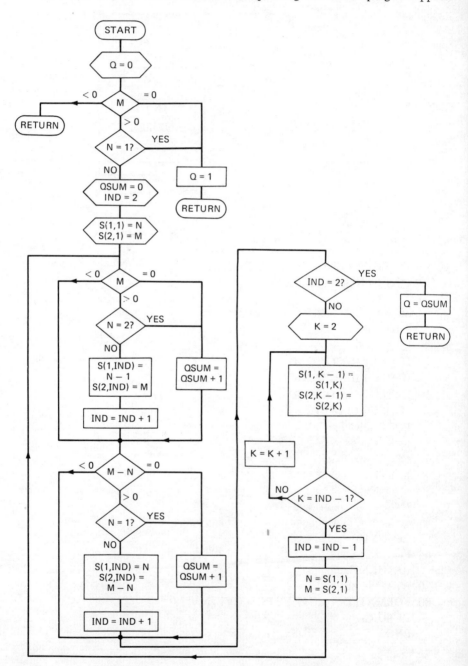

```
      INTEGER FUNCTION Q(MA, NA)
      DIMENSION IS (2,500)
      INTEGER QSUM
      M = MA
      N = NA
      Q = 0
      IF (M)999,10,30
   10 Q = 1
      RETURN
   30 IF (N.EQ.1) GO TO 10
      QSUM = 0
      IND = 2
      IS (1,1) = N
      IS (2,1) = M
   35 IF (M)40,50,60
   50 QSUM = QSUM+1
      GO TO 40
   60 IF (N.EQ.2) GO TO 50
      IS (1,IND) = N-1
      IS (2,IND) = M
      IND = IND+1
   40 IF (M-N)140,150,160
  150 QSUM = QSUM+1
      GO TO 140
  160 IF (N.EQ.1) GO TO 150
      IS (1,IND) = N
      IS (2,IND) = M-N
      IND = IND+1
  140 IF (IND.EQ.2) GO TO 98
      IIND = IND-1
      DO 200 K = 2, IIND
      IS (1,K-1) = IS (1,K)
      IS (2,K-1) = IS (2,K)
  200 CONTINUE
      IND = IND-1
      N = IS (1,1)
      M = IS (2,1)
      GO TO 35
   98 Q = QSUM
      RETURN
  999 STOP
      END
```

5.3 Your program probably has at least eighty source statements. As you can readily verify, the corresponding ALGOL program has around fifteen statements, which shows how much work can be saved by choosing the appropriate language for a particular job.

Chapter 7

7.1 *Algorithms:*
    (1)   A method to find the prime factors of a given integer.
    (2)   An optimal program for playing the game of noughts-and-crosses.
    (3)   An algorithm for driving from point A to point B.

Procedures:
    (1)   A procedure to become a millionaire. Stand on any busy street corner. Stop passers-by and ask for a million dollars.
    (2)   A procedure to prove that an arbitrary computer program is correct.

Heuristics:
    (1)   A heuristic to play the game of 'go'.

7.2 Let 2 be the second prime, 3 the third prime and so on.

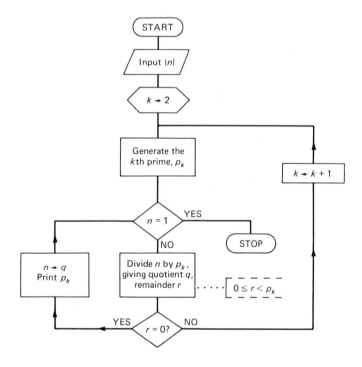

7.3 Use the facts that (1) the number of prime factors $\leqslant \log_2 n$, (2) the maximum prime factor is $|n|$.

Chapter 8

8.1   SUBROUTINE PERMS(NP,M,N,NC)
    C   GIVEN A PERMUTATION OF THE FIRST M INTEGERS
    C   STORED IN ARRAY NP(I), I=1,M A UNIQUE INTEGER
    C   N IS RETURNED FOR EACH OF THE M! PERMUTATIONS.
    C   1≤N≤M! THE PERMUTATION IS DESTROYED.
            DIMENSION NP(M),NC(M)
            NR = M
      15 K = 1
            MAX = NP(1)
            DO 20 I = 2,NR
                IF (MAX.GT.NP(I)) GO TO 20
                K = I
                MAX = NP(I)
      20 CONTINUE
            NC(NR) = K−1
            IF (NR.EQ.2) GO TO 100
            NSAVE = NP(NR)
            NP(NR) = NP(K)
            NP(K) = NSAVE
            NR = NR−1
            GO TO 15
     100 N = 1
            NPROD = 1
            MM = M−1
            DO 120 II = 1,MM
            I = II−1
            N = N+NC(M−I)*NPROD
            NPROD = 1
            DO 120 J = 1,II
            NPROD = NPROD*(M+1−J)
     120 CONTINUE
            RETURN
            END

8.2   It works in every case.

8.3 $\displaystyle\sum_{k=2}^{n} \{1 + \lceil \log_2 (k-1) \rceil \} = 1+2+3+3+4+4+4+4+ \ldots +q+q+q \ldots +q =$

$\underbrace{\quad}_{\substack{2 \\ \text{terms}}} \underbrace{\quad}_{\substack{4 \\ \text{terms}}} \underbrace{\quad\quad}_{\leqslant 2^{q-2} \text{ terms}}$

$$\leqslant 1 + \sum_{p=2}^{q} p 2^{p-2},$$

where $q = \lceil \log_2(n-1) \rceil + 1$.

$$\text{Now } \sum_{p=2}^{q} p2^{p-2} = \frac{1}{2} \frac{d}{dx} \sum_{p=2}^{q} x^p \bigg|_{x=2} = \frac{1}{2} \frac{d}{dx} \left(\frac{x^{q+1} - x^2}{x-1} \right) \bigg|_{x=2}$$

$$= (q-1)2^{q-1}$$

from which the desired result immediately follows.

Chapter 9

9.1   Mean $= E\{X\}$, variance $= E\{X^2\}-(E\{X\})^2$.

(a) Poisson distribution:

$$\text{mean} = \sum_{x=0}^{\infty} \frac{x e^{-\lambda}\lambda^x}{x!} = \lambda \sum_{x=0}^{\infty} \frac{e^{-\lambda}\lambda^x}{x!} = \lambda$$

$$E\{x^2\} = \sum_{x=0}^{\infty} \frac{x(x-1)e^{-\lambda}\lambda^x}{x!} + \sum_{x=0}^{\infty} \frac{x e^{-\lambda}\lambda^x}{x!} = \lambda^2 \sum_{x=0}^{\infty} \frac{e^{-\lambda}\lambda^x}{x!} + \lambda$$

$$= \lambda^2 + \lambda$$

$\therefore$ variance $= \lambda^2 + \lambda - \lambda^2 = \lambda$

(b) Uniform distribution:

$$\text{mean} = \int_{-\infty}^{\infty} X p(X) dX = \int_{a}^{b} \frac{X}{b-a}\, dX = \frac{X^2}{2(b-a)}\bigg|_{a}^{b} = \frac{b^2 - a^2}{2(b-a)} = \frac{b+a}{2}$$

$$E\{X^2\} = \int_{a}^{b} \frac{X^2}{b-a}\, dX = \frac{b^3 - a^3}{3(b-a)} = \frac{b^2 + ab + a^2}{3}$$

$$\therefore \text{variance} = \frac{b^2 + ab + a^2}{3} - \frac{(b+a)^2}{4} = \frac{(b-a)^2}{12}$$

(c) Normal distribution:

$$\text{mean} = \frac{1}{\sigma\sqrt{2\pi}} \int_{-\infty}^{\infty} t e^{-(t-\mu)^2/2\sigma^2}\, dt = \frac{1}{\sqrt{2\pi}} \int_{-\infty}^{\infty} (\sigma X + \mu)e^{-X^2/2} dX,$$

where $X = \dfrac{t-\mu}{\sigma}$

$$\therefore \text{mean} = \int_{-\infty}^{\infty} \frac{\mu}{\sqrt{2\pi}} X e^{-X^2/2} dX = \mu$$

Similarly, variance $= \dfrac{1}{\sqrt{2\pi}} \displaystyle\int_{-\infty}^{\infty} (\sigma X + \mu)^2 e^{-X^2/2} dX$, and we can set $\mu = 0$, since it is just a displacement of the y axis, and does not affect the variance

$$= \frac{\sigma^2}{\sqrt{2\pi}} \int_{-\infty}^{\infty} X^2 e^{-X^2/2} = \frac{\sigma^2}{\sqrt{2\pi}} \left\{ -Xe^{-X^2/2} \Big|_{-\infty}^{\infty} + \int_{-\infty}^{\infty} e^{-X^2/2} dX \right\}$$

$$= \frac{\sigma^2}{\sqrt{2\pi}} \left\{ \int_{-\infty}^{\infty} e^{-X^2/2} dX \right\} = \frac{\sigma^2}{\sqrt{2\pi}} \cdot \sqrt{2\pi} = \sigma^2$$

(integration by parts)

(d) Exponential distribution:

$$\text{mean} = \int_{0}^{\infty} \lambda X e^{-\lambda X} dX = -Xe^{-\lambda X} \Big|_{0}^{\infty} + \int_{0}^{\infty} e^{-\lambda X} dX \quad \text{(integration by parts)}$$

$$E\{X^2\} = \int_{0}^{\infty} \lambda X^2 e^{-\lambda X} = -X^2 e^{-\lambda X} \Big|_{0}^{\infty} + \int_{0}^{\infty} 2Xe^{-\lambda X} dX = \frac{2}{\lambda^2}$$

$$\text{variance} = \frac{2}{\lambda^2} - \frac{1}{\lambda^2} = \frac{1}{\lambda^2}$$

Bibliography

Chapter 1

Dahl, O. J., Dijkstra, E. W. and Hoare, C. A., *Structured Programming*, (New York: Academic Press, 1972).

Farina, M. V., *Flowcharting*, (New Jersey: Prentice-Hall, 1970).

Hughes, M. L., Shank, R. M., and Stein, E. S., *Decision Tables*, (Wayne, Pennsylvania: MDI Publications, 1968).

Humby, E., *Programs from Decision Tables*, (London and New York: MacDonald/ Elsevier, 1973).

Kernighan, B. W. and Plauger, P. J., *The Elements of Programming Style*, (New York: McGraw-Hill, 1974).

Pollack, S. L., Hicks, H. T. and Harrison, W. J., *Decision Tables: Theory and Practice*, (New York: J. Wiley & Sons, 1971).

Pooch, U. W., 'Translation of decision tables', *Comp. Surveys*, **6**, 127, 1974.

Schriber, T. J., *Fundamentals of Flowcharting*, (New York: J. Wiley & Sons, 1969).

Wirth, N., *Systematic Programming*, (New Jersey: Prentice-Hall, 1973).

Chapter 2

BASIC, ICL Technical Publication 4953, (Letchworth, Herts: ICL Printing Services, 1972).

Kemeny, J. G., and Kurtz, T. E., *BASIC (User's Manual)*, (Hanover, N.H.: Dartmouth College Computation Centre, 1966).

Kemeny, J. G., and Kurtz, T. E., *BASIC Programming*, (New York: J. Wiley & Sons, 1967).

Sharpe, W. F., *BASIC: An Introduction to Computer Programming Using the BASIC Language*, (New York: The FREE PRESS, London: Collier-Macmillan, 1967).

Chapter 3

Blatt, J. M., *Introduction to FORTRAN IV Programming,* (California: Goodyear Pub. Co., 1967).

McCracken, D., *A Guide to FORTRAN IV Programming*, (New York: J. Wiley & Sons Inc., 1965).

Sammet, Jean E., *Programming Languages: History and Fundamentals*, (New Jersey: Prentice-Hall, Inc. 1969).

Watters, J., *FORTRAN Programming,* (London: Heinemann Educational Books, 1970).

U.S.A. Standard FORTRAN, U.S.A. Standards Institute, USAS X3–9–1966, New York, March 1966.

Chapter 4

ALGOL: Language; Technical Publication 3340 (Letchworth, Herts: ICL, 1965).

Bottenbruch, H., 'Structure and Use of ALGOL 60', *Journal of the A.C.M.,* 9, 161–221, 1962.

Naur, P., (Editor), 'Revised report on the Algorithmic Language ALGOL 60', [In almost any ALGOL text]

Sammet, Jean E., *Programming Languages: History and Fundamentals*, (New Jersey: Prentice-Hall Inc., 1969).

Report on Input-Output Procedures for ALGOL 60 (IFIP)', *Comm. A.C.M.,* 7, 628–630 1964.

Chapter 5

Barron, D. W., *Recursive Techniques in Programming*, (London & New York: MacDonald-Elsevier, 1968).

Day, A. C., *FORTRAN Techniques: with special reference to non-numerical applications,* (Cambridge: Cambridge University Press, 1972).

Guttmann, A. J., 'Programming Recursively Defined Functions in FORTRAN', *Intl. Jour. Comp. & Info. Sciences,* Spring 1976.

Higman, B., *A Comparative Study of Programming Languages*, (London & New York: MacDonald-Elsevier, 1967).

McMahon, J. T., '*ALGOL* vs *FORTRAN*', *Datamation,* April 1972, pp. 88–89.

Morris, J., 'Programming Recursive Functions in FORTRAN', *Software Age,* January 1969, pp. 38–42.

Sammet, Jean E., *Programming Languages: History and Fundamentals*, (New Jersey: Prentice-Hall Inc., 1969).

Sanders, N., and Fitzpatrick, C., 'ALGOL & FORTRAN revisited', *Datamation,* January 1963, pp. 30–32.

Siles, R. L., *ALGOL-W Reference Manual,* (Computing Centre, University of Waterloo, 1972).

Wirth, N., and Hoare, C. A. R. 'A Contribution to the Development of ALGOL', *Comm A.C.M.,* 9, 1966, 413.

Chapter 6

Blatt, J. M., *Introduction to FORTRAN IV programming*, (California: Goodyear, 1967).

Brown, A. R., and Sampson, W. A., *Program Debugging*, (Amsterdam: Elsevier, 1973).

Kernighan, B. W., and Plauger, P. J., *The Elements of Programming Style*, (New York: McGraw-Hill, 1974).

Kreitzberg, C. B., and Shneiderman, B., *The elements of FORTRAN style*, (New York: Harcourt, Brace & Johvanovich Inc., 1972).

Larson, C., 'The efficient use of FORTRAN', *Datamation,* August 1, 1971, pp. 24–31.

Ralston, A., *A First Course in Numerical Analysis*, (New York: McGraw-Hill, 1965).

Chapter 7

Aho, A. V., Hopcroft, J. E., and Ullman, J. D., *The Design and Analysis of Computer Algorithms,* (Massachusetts: Addison-Wesley, 1974).

Knuth, D., *The Art of Computer Programming,* Vol. 1: Fundamental Algorithms, (Massachusetts: Addison-Wesley, 1968).

Stein, J., 'Computational problems associated with Racah algebra' *J. Comp. Phys.,* 1, 1967, 397–405.

Chapter 8

Berztiss, A. T., *Data Structures: Theory and Practice,* (New York: Academic Press, 1971).

Flores, I., 'Computer Time for address calculation sorting', *Journal of the A.C.M.,* I 1960, pp. 389–409.

Gotlieb, C. S., 'Sorting on Computers', in *Applications of Digital Computers*, pp.68–8 (Boston, Ginn & Co., 1963).

Knott, G. D., 'Hashing Functions', *The Computer Journal,* 18, 1975, 265.

Knuth, D. E., *Semi-numerical Algorithms,* (Massachusetts: Addison-Wesley, 1969).

Knuth, D. E., *Sorting and Searching,* (Massachusetts: Addison-Wesley, 1973).

Maurer, W. D., *Programming – An Introduction to Computer Techniques,* (San Francisco, Holden-Day Inc., 1972).

Maurer, W. D., and Lewis, T. G., 'Hash Table Methods', *A.C.M. Computing Surveys,* 7, 1975, 5.

Shell, D. L., 'A high-speed sorting procedure', *Comm. A.C.M.,* 2, 1959, pp. 30–32.

:hapter 9

bramowitz, M., and Stegun, I. A., Eds, *Handbook of Mathematical Functions*, NBS, Washington, 1964.

ɔx, G. E. P. and Muller, M. G., 'A Note on the Generation of Random Normal Deviates', *Ann. Math. Stat., 29*, 1958, 610.

ɔveyou, R. R. and MacPherson, R. D., 'Fourier Analysis of Uniform Random Number Generators', *JACM, 14*, 1967, 100–19.

ammersley, J. M. and Hanscomb, D. C., *Monte Carlo Methods,* (London: Methuen, 1964).

astings, C., Jr., *Approximations for Digital Computers,* (Princeton, N.J.: Princeton University Press, 1955).

emmerle, W. J., *Statistical Computations on a Digital Computer,* (Massachusetts: Blaisdell, 1967).

nsson, B., *Random Number Generators,* (Stockholm: V. Pettersons, 1966).

ɪuth, D., 'Semi-numerical Algorithms', in *The Art of Computer Programming,* Vol. 2, (Massachusetts: Addison-Wesley, 1969).

ɛhmer, D. H., Mathematical Methods in Large Scale Computing Units', *Annals. Comp. Lab. Harvard University, 26*, 1951, 141–6.

acLaren, M. D., and Marsaglia, G., 'Uniform Random Generators', *JACM, 12*, 1965, 83–89.

arsaglia, G., 'Improving the Polar Method for Generating a Pair of Normal Random Deviates', Boeing Scientific Res. Labs. Doc. D1–82–0203, Sept., 1962.

arsaglia, G., and MacLaren, M. D., 'A Fast Procedure for Generating Normal Random Variables', *CACM, 7*, 1964, 4–10.

aylor, T. H., Balintfy, J. L., Burdick, D. S., Chu, K., *Computer Simulation Techniques,* (New York: Wiley, 1966).

ɪn Neumann, J., *Collected Works, 5*, (New York: Macmillan, 1963).

Index

Programming and Algorithms discusses and develops the contemporary view of computer programming and algorithm design. Unlike most other introductory programming texts, some prior knowledge and experience with a programming language is assumed. This approach recognizes a growing educational trend in which students receive some elementary programming instruction at a very early stage – often as part of a first year university or similar Mathematics course, or even at school.

The first chapter develops the ideas of top-down program design, structured programming, and program style. Six basic control structures are introduced, and their role in writing well-styled, well-structured programs is fully discussed. The three languages used to represent a program in the planning stage ; natural language, decision tables, and flowcharts, are also introduced at this stage.
The next three chapters are devoted to a discussion of the three most commonly used scientific programming languages : BASIC, FORTRAN, and ALGOL.

The two languages FORTRAN and ALGOL are then discussed and compared and this comparison leads to a discussion of recursive procedures and a way of simulating these in FORTRAN is presented. The question of program planning and design forms the subject of the next chapter, with special emphasis being placed on the most efficient methods of coding various common problems. The following chapter considers the subject of algorithm design and analysis.

The remainder of the book continues the study of algorithms by considering a number of numerical and non-numerical problems.

At the end of each chapter a number of problems are given, most of which are designed to illustrate the material in that chapter while a few are chosen to extend it. Solutions to most of the problems are given at the end of the book. A bibliography is also given at the end of the book in which pertinent references for each chapter can be readily found.

This book is based on an introductory programming course of thirty lectures that has been given at the University of Newcastle, New South Wales, Australia for the past five years.

ISBN 0 435 77541 3

£3.50 net

Heinemann Educational Books